Probability Distributions Used in Reliability Engineering

Probability Distributions
Used in
Reliability Engineering

Andrew N. O'Connor
Mohammad Modarres
Ali Mosleh

Published by the Andrew N. O'Connor, Mohammad Modarres, and Ali Mosleh

International Standard Book Number (ISBN): 978-0-9966468-1-9

In memory of Willie Mae Webb

This book is dedicated to the memory of Miss Willie Webb who passed away on April 10 2007 while working at the Center for Risk and Reliability at the University of Maryland (UMD). She initiated the concept of this book, as an aid for students conducting studies in Reliability Engineering at the University of Maryland. Upon passing, Willie bequeathed her belongings to fund a scholarship providing financial support to Reliability Engineering students at UMD.

Preface

Reliability Engineers are required to combine a practical understanding of science and engineering with probability and statistics. The reliability engineer's understanding of probability and statistics is focused on the practical application of a wide variety of accepted probabilistic methods. Most reliability texts provide only a basic introduction to probability distributions or only provide a detailed reference to few distributions. Most textbooks in probability and statistics provide theoretical detail which is outside the scope of likely reliability engineering tasks. As such the objective of this book is to provide a single reference text of closed form statistical and probabilistic formulas and approximations used in reliability engineering.

This book provides details on many probability distributions used in reliability engineering. Each distribution section provides a graphical visualization and formulas for distribution parameters, along with reliability functions and other related formulas. Common statistics such as moments and percentile formulas are followed by likelihood functions and in many cases the derivation of maximum likelihood estimates. Bayesian non-informative and conjugate priors are provided followed by a discussion on the distribution characteristics and applications in reliability engineering. The book includes many numerical examples showing applications of the distribution function in the context of reliability engineering problems. Each section concludes with online and hardcopy references which can provide further information followed by the relationship to other distributions.

The book is divided into six parts. Part 1 provides a brief coverage of the fundamentals of probability distributions within a reliability engineering context. Part 1 is limited to concise explanations aimed to familiarize readers. For further understanding the reader is referred to the relevant references. Part 2 to Part 6 cover Common Life Distributions, Bathtub Life Distributions, Univariate Continuous Distributions, Univariate Discrete Distributions and Multivariate Distributions, respectively.

The authors would like to thank the many students in the Reliability Engineering Program who have commented on the early versions of this book, but particularly Dr. Reuel Smith's contributions and proof reading were significant.

Contents

Contents   iii

1. Fundamentals of Probability Distributions

1.1. Probability Theory

1.1.1. Theory of Probability

The theory of probability formalizes the representation of probabilistic concepts through a set of rules. The most common reference to formalizing the rules of probability is through a set of *axioms* proposed by Kolmogorov in 1933. Considering E_i as an event in the event space $\Omega = \cup_{i=1}^{n} E_i$ with n different events, the axioms are:

$$0 \leq P(E_i) \leq 1$$

$$P(\Omega) = 1 \text{ and } P(\phi) = 0$$

$$P(E_1 \cup E_2 \cup \cdots \cup E_n) = \sum_i P(E_i),$$

when events E_i are mutually exclusive.

Other representations of uncertainty exist such as fuzzy logic and theory of evidence (Dempster-Shafer model) which do not follow the theory of probability but almost all reliability concepts are defined based on probability as the metric of uncertainty. For a justification of probability theory see (Singpurwalla 2006).

1.1.2. Interpretations of Probability

The two most common interpretations of probability are:

- **Frequency Interpretation.** In the frequentist interpretation of probability, the probability of an event (failure) is defined as:

$$P(K) = \lim_{n \to \infty} \frac{n_f}{n}$$

  Also known as the classical representation, this interpretation assumes there exists a real probability of an event, p. The analyst uses the observed frequency of the event to estimate the value of p. The more historic events that have occurred, the more confident the analyst is of the estimation of p. This approach does have limitations, for instance when data from events are not available (e.g. no failures occur in a test) p cannot be estimated and this method cannot incorporate other information such a "soft evidence" in form of expert opinion.

- **Subjective Interpretation.** The subjective interpretation of probability is also known as the Bayesian school of thought. This interpretation views the probability of an event as degree of belief the analyst has on the occurrence of event. This means probability is a product of the analyst's state of knowledge. Any evidence which would change the analyst's degree of belief must be considered when estimating the probability (including soft evidence). The assumption is made that the probability assessment is made by a rational person, where any other rational person having the same state of knowledge would make the same assessment.

The subjective interpretation has the flexibility of including many types of evidence to assist in estimating the probability of an event. This is important in many reliability applications where the events of interest (e. g, system failure) are rare.

1.1.3. Laws of Probability

The following rules of logic form the basis for many mathematical operations within the theory of probability.

Let $X = E_i$ and $Y = E_j$ be two events within the sample space Ω where $i \neq j$.

Boolean Laws of probability are (Modarres et al. 1999, p.25):

$X \cup Y = Y \cup X$ $X \cap Y = Y \cap X$	Commutative Law
$X \cup (Y \cup Z) = (X \cup Y) \cup Z$ $X \cap (Y \cap Z) = (X \cap Y) \cap Z$	Associative Law
$X \cap (Y \cup Z) = (X \cap Y) \cup (X \cap Z)$	Distributive Law
$X \cup X = X$ $X \cap X = X$	Idempotent Law
$X \cup \bar{X} = \Omega$ $X \cap \bar{X} = \emptyset$ $\bar{\bar{X}} = X$	Complementation Law
$\overline{(X \cup Y)} = \bar{X} \cap \bar{Y}$ $\overline{(X \cap Y)} = \bar{X} \cup \bar{Y}$	De Morgan's Theorem

Two events are mutually exclusive (or disjoint) if:
$$X \cap Y = \emptyset, \qquad P(X \cap Y) = 0$$

Two events are independent if one event Y occurring does not affect the probability of the second event X occurring:

$$P(X|Y) = P(X)$$

The rules for evaluating the probability of compound events are:

Addition Rule:
$$P(X \cup Y) = P(X) + P(Y) - P(X \cap Y)$$
$$= P(X) + P(Y) - P(X) P(Y|X)$$

Multiplication Rule:
$$P(X \cap Y) = P(X) P(Y|X) = P(Y) P(X|Y)$$

When X and Y are independent:
$$P(X \cup Y) = P(X) + P(Y) - P(X) P(Y)$$
$$P(X \cap Y) = P(X) P(Y)$$

Generalizations of these rules:

$$P(E_1 \cup E_2 \cup ... \cup E_n) = [P(E_1) + P(E_2) + \cdots + P(E_n)]$$
$$- [P(E_1 \cap E_2) + P(E_1 \cap E_3) + \cdots + P(E_{n-1} \cap E_n)]$$
$$+ [P(E_1 \cap E_2 \cap E_3) + P(E_1 \cap E_2 \cap E_4) + \cdots]$$
$$- \cdots (-1)^{n+1}[P(E_1 \cap E_2 \cap ... \cap E_n)]$$

$$P(E_1 \cap E_2 \cap ... \cap E_n) = P(E_1) . P(E_2|E_1). P(E_3|E_1 \cap E_2)$$
$$... P(E_n|E_1 \cap E_2 \cap ... \cap E_{n-1})$$

1.1.4. Law of Total Probability

The probability of X can be obtained by the following summation;

$$P(X) = \sum_{i=1}^{n_A} P(A_i)P(X|A_i)$$

where $A = \{A_1, A_2, ..., A_{n_A}\}$ is a partition of the sample space, Ω, and all the elements of A are mutually exclusive, $A_i \cap A_j = \emptyset$, and the union of all A elements cover the complete sample space, $\cup_{i=1}^{n_A} A_i = \Omega$.

For example:

$$P(X) = P(X \cap Y) + P(X \cap \bar{Y})$$
$$= P(Y)P(X|Y) + P(Y)P(X|\bar{Y})$$

1.1.5. Bayes Theorem

Bayes Theorem, can be derived from the multiplication rule and the law of total probability as follows:

$$P(\theta) P(E|\theta) = P(E) P(\theta|E)$$

$$P(\theta|E) = \frac{P(\theta) P(E|\theta)}{P(E)}$$

$$P(\theta|E) = \frac{P(\theta) P(E|\theta)}{\sum_i P(E|\theta_i)P(\theta_i)}$$

θ the unknown of interest (UOI).

E the observed random variable, evidence (data).

$P(\theta)$ the prior state of knowledge about θ without the evidence. Also denoted as $\pi_o(\theta)$.

$P(E|\theta)$ the likelihood of observing the evidence given the UOI. Also denoted as $L(E|\theta)$.

$P(\theta|E)$ the posterior state of knowledge about θ given the evidence. Also denoted as $\pi(\theta|E)$.

$\sum_i P(E|\theta_i)P(\theta)$ is the normalizing constant.

Thus, Bayes theorem enables us to use the evidence, E, to make inference about the unobserved event or parameter θ .

The continuous form of Bayes' Law can be written as:
$$\pi(\theta|E) = \frac{\pi_o(\theta)\,L(E|\theta)}{\int \pi_o(\theta)\,L(E|\theta)\,d\theta}$$
In Bayesian statistics the state of knowledge (uncertainty) of an unknown of interest is quantified by assigning a probability distribution to its possible values. Bayes theorem provides a mathematical means by which this uncertainty can be updated given new evidence.

1.1.6. Likelihood Functions

In the frequentist inference, the *likelihood function* is the probability of observing the evidence (e.g., a sample of i.i.d.[1] data that come from a parent population), E, given the probability distribution with parameters, $\boldsymbol{\theta}$. The probability of observing events is the product of each event likelihood:
$$L(\boldsymbol{\theta}|E) = c\prod_i L(\boldsymbol{\theta}|t_i)$$
where, c is a combinatorial constant which quantifies the number of combinations which the observed evidence could have occurred. Methods which use the likelihood function in parameter estimation do not depend on the constant and so it is omitted.

The following table summarizes the likelihood functions for different types of observations:

Table 1: Summary of Likelihood Functions (Klein & Moeschberger 2003, p.74)

Type of Observation	Likelihood Function	Example Description
Exact Lifetimes	$L_i(\theta|t_i) = f(t_i|\theta)$	Failure time is known
Right Censored	$L_i(\theta|t_i) = R(t_i|\theta)$	Component survived to time t_i
Left Censored	$L_i(\theta|t_i) = F(t_i|\theta)$	Component failed before time t_i
Interval Censored	$L_i(\theta|t_i) = F(t_i^{RI}|\theta) - F(t_i^{LI}|\theta)$	Component failed between t_i^{LI} and t_i^{RI}
Left Truncated	$L_i(\theta|t_i) = \dfrac{f(t_i|\theta)}{R(t_L|\theta)}$	Component failed at time t_i where observations are truncated before t_L.
Right Truncated	$L_i(\theta|t_i) = \dfrac{f(t_i|\theta)}{F(t_U|\theta)}$	Component failed at time t_i where observations are truncated after t_U.

---

[1] Identically and Independently Distributed

Interval Truncated	$$L_i(\theta|t_i) = \frac{f(t_i|\theta)}{F(t_U|\theta) - F(t_L|\theta)}$$	Component failed at time t_i where observations are truncated before t_L and after t_U.

The Likelihood function is used in Bayesian inference and maximum likelihood parameter estimation techniques. In both instances any constant in front of the likelihood function becomes irrelevant. Such constants are therefore not included in the likelihood functions given in this book (nor in most references).

For example, consider the case where a test is conducted on n components with an exponential time to failure distribution. The test is terminated at time t_s during which r components failed at times $t_1, t_2, \dots, t_r$ and $s = n - r$ components survived. Using the exponential distribution to construct the likelihood function we obtain:

$$L(\lambda|E) = \prod_{i=1}^{n_F} f(\lambda|t_i^F) \prod_{i=1}^{n_S} R(\lambda|t_i^S)$$

$$= \prod_{i=1}^{n_F} \lambda e^{-\lambda t_i^F} \prod_{i=1}^{n_S} e^{-\lambda t_i^S}$$

$$= \lambda^{n_F} e^{-\lambda \sum_{i=1}^{n_F} t_i^F} e^{-\lambda \sum_{i=1}^{n_S} t_i^S}$$

$$= \lambda^{n_F} e^{-\lambda\left(\sum_{i=1}^{n_F} t_i^F + \sum_{i=1}^{n_S} t_i^S\right)}$$

Alternatively, because the test described is a homogeneous Poisson process[2] the likelihood function could also have been constructed using a Poisson distribution. The data can be stated as seeing r failure in time t_T where t_T is the total time on test $t_T = \sum_{i=1}^{n_F} t_i^F + \sum_{i=1}^{n_S} t_i^S$. Therefore, the likelihood function would be:

$$L(\lambda|E) = f(\lambda|n_F, t_T)$$

$$= \frac{(\lambda t_T)^{n_F}}{n_F!} e^{-\lambda t_T}$$

$$= c\lambda^{n_F} e^{-\lambda t_T}$$

$$= \lambda^{n_F} e^{-\lambda\left(\sum_{i=1}^{n_F} t_i^F + \sum_{i=1}^{n_S} t_i^S\right)}$$

As mentioned earlier, in estimation procedures within this book, the constant c can be ignored. As such, the two likelihood functions are equal. For more information see (Meeker & Escobar 1998, p.36) or (Rinne 2008, p.403).

---

[2] Homogeneous in time, where it does not matter if you have n components on test at once (exponential test), or you have a single component on test which is replaced after failure n times (Poisson process), the evidence produced will be the same.

1.1.7. Fisher Information Matrix

The Fisher Information Matrix has many uses but in reliability applications it is most often used to estimated confidence intervals of probability distribution parameters and create Jeffery's non-informative priors in Bayesian estimation. There are two types of Fisher information matrices, the Expected Fisher Information Matrix $I(\theta)$, and the Observed Fisher Information Matrix $J(\theta)$.

The *Expected Fisher Information Matrix* is obtained from a log-likelihood function from a single random variable. The random variable is replaced by its expected value.

For a single parameter distribution:

$$I(\theta) = -E\left[\frac{\partial^2 \Lambda(\theta|x)}{\partial \theta^2}\right] = \left[\left(\frac{\partial \Lambda(\theta|x)}{\partial \theta}\right)^2\right]$$

where Λ is the log-likelihood function, x is a sample i.i.d. data, and $E[U] = \int U f(x)dx$. For a distribution with p parameters the *Expected Fisher Information Matrix* is:

$$I(\theta) = \begin{bmatrix} -E\left[\frac{\partial^2 \Lambda(\theta|x)}{\partial \theta_1^2}\right] & -E\left[\frac{\partial^2 \Lambda(\theta|x)}{\partial \theta_1 \partial \theta_2}\right] & \cdots & -E\left[\frac{\partial^2 \Lambda(\theta|x)}{\partial \theta_1 \partial \theta_p}\right] \\ -E\left[\frac{\partial^2 \Lambda(\theta|x)}{\partial \theta_2 \partial \theta_1}\right] & -E\left[\frac{\partial^2 \Lambda(\theta|x)}{\partial \theta_2^2}\right] & \cdots & -E\left[\frac{\partial^2 \Lambda(\theta|x)}{\partial \theta_2 \partial \theta_p}\right] \\ \vdots & \vdots & \ddots & \vdots \\ -E\left[\frac{\partial^2 \Lambda(\theta|x)}{\partial \theta_p \partial \theta_1}\right] & -E\left[\frac{\partial^2 \Lambda(\theta|x)}{\partial \theta_p \partial \theta_2}\right] & \cdots & -E\left[\frac{\partial^2 \Lambda(\theta|x)}{\partial \theta_p^2}\right] \end{bmatrix}$$

The *Observed Fisher Information Matrix* is obtained from a likelihood function constructed from n observed samples from the distribution. The *expectation* term is dropped.

For a single parameter distribution:

$$J_n(\theta) = -\sum_{i=1}^{n} \frac{\partial^2 \Lambda(\theta|x_i)}{\partial \theta^2}$$

For a distribution with p parameters the Observed Fisher Information Matrix is:

$$J_n(\theta) = \sum_{i=1}^{n} \begin{bmatrix} -\frac{\partial^2 \Lambda(\theta|x_i)}{\partial \theta_1^2} & -\frac{\partial^2 \Lambda(\theta|x_i)}{\partial \theta_1 \partial \theta_2} & \cdots & -\frac{\partial^2 \Lambda(\theta|x_i)}{\partial \theta_1 \partial \theta_p} \\ -\frac{\partial^2 \Lambda(\theta|x_i)}{\partial \theta_2 \partial \theta_1} & -\frac{\partial^2 \Lambda(\theta|x_i)}{\partial \theta_2^2} & \cdots & -\frac{\partial^2 \Lambda(\theta|x_i)}{\partial \theta_2 \partial \theta_p} \\ \vdots & \vdots & \ddots & \vdots \\ -\frac{\partial^2 \Lambda(\theta|x_i)}{\partial \theta_p \partial \theta_1} & -\frac{\partial^2 \Lambda(\theta|x_i)}{\partial \theta_p \partial \theta_2} & \cdots & -\frac{\partial^2 \Lambda(\theta|x_i)}{\partial \theta_p^2} \end{bmatrix}$$

It can be seen that as n becomes large, the average value of the random variable approaches its expected value and so the following asymptotic relationship exists between the observed and expected Fisher information matrices:

$$\operatorname*{plim}_{n \to \infty} \frac{1}{n} J_n(\boldsymbol{\theta}) = I(\boldsymbol{\theta})$$

For large n the following approximation can be used:

$$J_n \approx n I(\boldsymbol{\theta})$$

When evaluated at $\theta = \hat{\theta}$ the observed Fisher information matrix estimates the variance-covariance matrix:

$$V = \left[J_n(\boldsymbol{\theta} = \hat{\boldsymbol{\theta}}) \right]^{-1} = \begin{bmatrix} Var(\hat{\theta}_1) & Cov(\hat{\theta}_1, \hat{\theta}_2) & \cdots & Cov(\hat{\theta}_1, \hat{\theta}_d) \\ Cov(\hat{\theta}_1, \hat{\theta}_2) & Var(\hat{\theta}_2) & \cdots & Cov(\hat{\theta}_2, \hat{\theta}_d) \\ \vdots & \vdots & \ddots & \vdots \\ Cov(\hat{\theta}_1, \hat{\theta}_d) & Cov(\hat{\theta}_2, \hat{\theta}_d) & \cdots & Var(\hat{\theta}_d) \end{bmatrix}$$

1.2. Distribution Functions

1.2.1. Random Variables

Probability distributions are used to model random events for which the outcome is uncertain such as the time of failure for a component. Before placing a demand on that component, the time it will fail is unknown. The distribution of the probability of failure at different times is modeled by a probability distribution. In this book random variables will be denoted as capital letter such as T for time. When the random variable assumes a known value, we denote this by small caps such as t for time. For example, if we wish to find the probability that the component fails before time t_1, we would find $P(T \leq t_1)$.

Random variables are classified as either *discrete* or *continuous*. In a discrete distribution, the random variable can take on a distinct or countable number of possible values such as number of demands to failure. In a continuous distribution the random variable is not constrained to distinct possible values such as time-to-failure distribution.

This book will denote continuous random variables as X or T, and discrete random variables as K.

1.2.2. Distribution Parameters

The parameters of a distribution are the variables which need to be specified in order to completely describe the distribution. Often parameters are classified by the effect they have on the distributions. Shape parameters define the shape of the distribution, scale parameters stretch the distribution along the random variable axis, and location parameters shift the distribution along the random variable axis. The reader is cautioned that the parameters for a distribution may change depending on the text. Therefore, before using formulas from other sources the parameterization needs to be confirmed.

Understanding the effect of changing a distribution's parameter value can be a difficult task. At the beginning of each section a graph of the distribution is shown with varied parameters.

1.2.3. Probability Density Function

A probability density function (pdf), denoted as $f(t)$ is any function which is always positive and is normalized to a cumulative value of one over the entire range of its random variable:

$$\int_{-\infty}^{\infty} f(t)\, dt = 1, \qquad \sum_k p(k) = 1$$

The probability of an event occurring between limits a and b is the area under the pdf:

$$P(a \leq T \leq b) = \int_a^b f(t)\, dt = F(b) - F(a)$$

$$P(a \leq K \leq b) = \sum_{i=a}^b p(k) = F(b) - F(a)$$

The instantaneous value of a discrete distribution at k_i can be obtained by minimizing the limits to $[k_{i-1}, k_i]$:

$$P(K = k_i) = P(k_i < K \le k_i) = p(k)$$

The instantaneous value of a continuous pdf is infinitesimal. This result can be seen when minimizing the limits to $[t, t + \Delta t]$:

$$P(T = t) = \lim_{\Delta t \to 0} P(t < T \le t + \Delta t) = \lim_{\Delta t \to 0} f(t).\Delta t$$

Therefore, the reader must remember that in order to calculate the probability of an event, an interval for the random variable must be used. Furthermore, a common misunderstanding is that a pdf cannot have a value above one because the probability of an event occurring cannot be greater than one. As can be seen above this is true for discrete distributions, only because $\Delta k = 1$. However, for continuous the case the pdf is multiplied by a small interval Δt, which ensures that the probability an event occurring within the interval Δt is less than one.

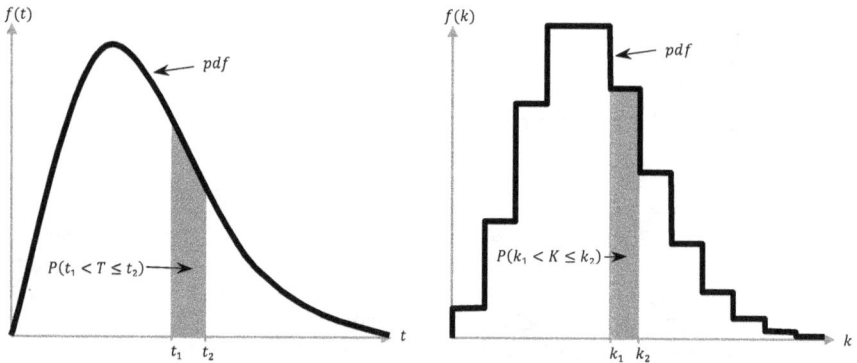

Figure 1: Left: continuous pdf, right: discrete pdf

To derive the continuous pdf relationship to the cumulative density function (cdf), $F(t)$:

$$\lim_{\Delta t \to 0} f(t).\Delta t = \lim_{\Delta t \to 0} P(t < T \le t + \Delta t) = \lim_{\Delta t \to 0} \{F(t + \Delta t) - F(t)\} = \lim_{\Delta t \to 0} \Delta F(t)$$

$$f(t) = \lim_{\Delta t \to 0} \frac{\Delta F(t)}{\Delta t} = \frac{dF(t)}{dt}$$

The shape of the pdf can be obtained by plotting a normalized histogram of an infinite number of samples from a distribution.

It should be noted when plotting a discrete pdf the points from each discrete value should not be joined. For ease of explanation using the area under the graph argument the step plot is intuitive but implies a non-integer random variable. Instead stem plots or column plots are often used.

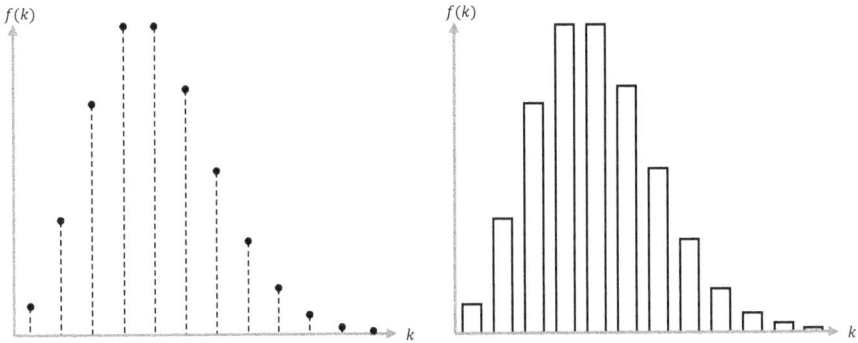

Figure 2: Discrete data plotting. Left stem plot. Right column plot.

1.2.4. Cumulative Distribution Function

The cumulative density function (cdf), denoted by $F(t)$ is the probability of the random event occurring before t, $P(T \leq t)$. For a discrete cdf the height of each step is the pdf value $f(k_i)$.

$$F(t) = P(T \leq t) = \int_{-\infty}^{t} f(x)\, dx, \qquad F(k) = P(K \leq k) = \sum_{k_i \leq k} f(k_i)$$

The limits of the cdf for $-\infty < t < \infty$ and $0 \leq k \leq \infty$ are given as:

$$\lim_{t \to -\infty} F(t) = 0, \qquad F(-1) = 0$$

$$\lim_{t \to \infty} F(t) = 1, \qquad \lim_{k \to \infty} F(k) = 1$$

The cdf can be used to find the probability of the random even occurring between two limits:

$$P(a \leq T \leq b) = \int_{a}^{b} f(t)\, dt = F(b) - F(a)$$

$$P(a \leq K \leq b) = \sum_{i=a}^{b} f(k) = F(b) - F(a)$$

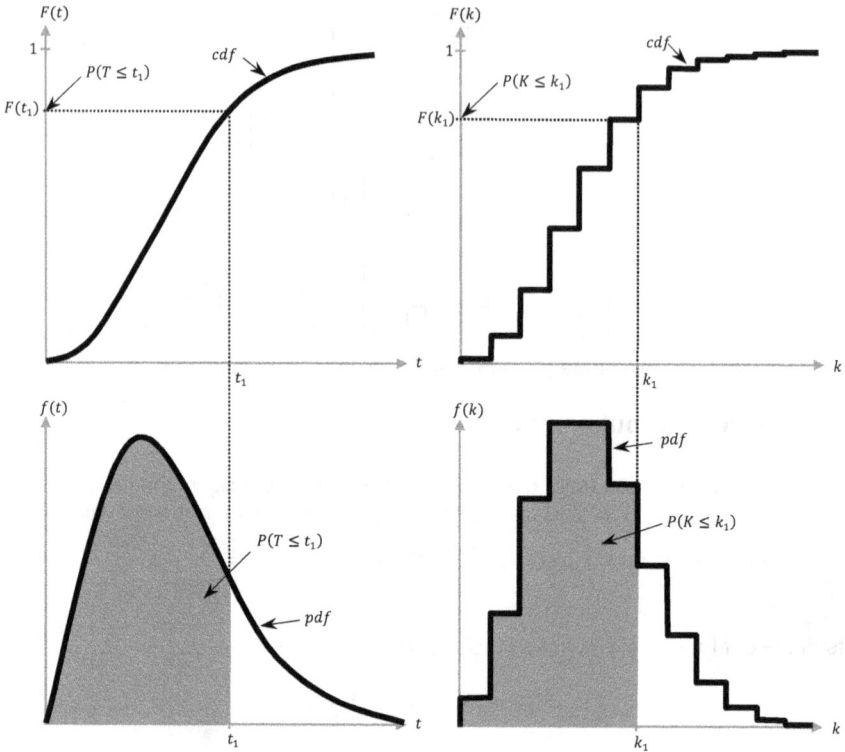

Figure 3: Left: continuous cdf/pdf, right: discrete cdf/pdf

1.2.5. Reliability Function

The *reliability function*, also known as the *survival function*, is denoted as $R(t)$. It is the probability that the random event (time-of-failure) occurs after t.

$$R(t) = P(T > t) = 1 - F(t), \qquad R(k) = P(T > k) = 1 - F(k)$$

$$R(t) = \int_{t}^{\infty} f(t) \, dt, \qquad R(k) = \sum_{i=k+1}^{\infty} f(k_i)$$

It should be noted that in most publications the discrete reliability function is defined as $R^*(k) = P(T \geq k) = \sum_{i=k}^{\infty} f(k)$. This definition results in $R^*(k) \neq 1 - F(k)$. Despite this problem it is the most common definition and is included in all the references in this book except (Xie, Gaudoin, et al. 2002)

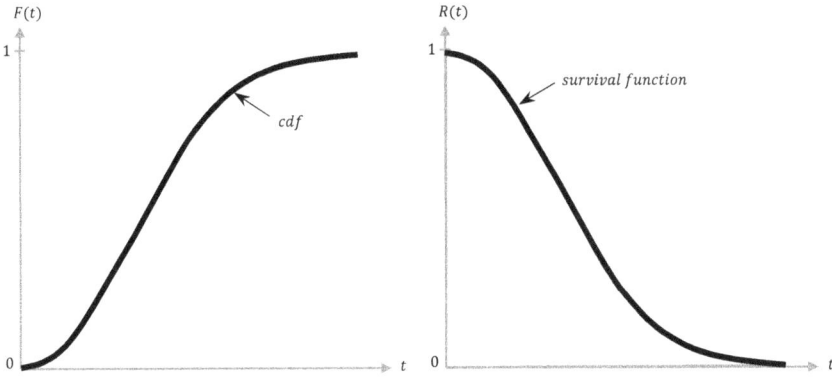

Figure 4: Left continuous cdf, right continuous survival function

1.2.6. Conditional Reliability Function

The *conditional reliability function*, denoted as $m(x)$ is the probability of the component surviving its mission given that it has survived to time t.

$$m(x) = R(x|t) = \frac{R(t+x)}{R(t)}$$

Where:
 t is the given time for which we know the component survived.
 x is new random variable defined as the time after t. $x = 0$ at t.

1.2.7. 100α% Percentile Function

The 100α% percentile function is the interval $[0, t_\alpha]$ for which the area under the pdf is α.

$$t_\alpha = F^{-1}(\alpha)$$

1.2.8. Mean Residual Life

The *mean residual life* (MRL), denoted as $u(t)$, is the expected life given the component has survived to time, t.

$$u(t) = \int_0^\infty R(x|t)\, dx = \frac{1}{R(t)} \int_t^\infty R(x)\, dx$$

1.2.9. Hazard Rate

The *hazard function*, denoted as h(t), is the conditional probability that a component fails in a short time interval, given that it has survived from time zero until the beginning of the time interval. For the continuous case the probability that an item will fail in a time interval given the item was functioning at time t is:

$$P(t < T < t + \Delta t | T > t) = \frac{P(t < T < t + \Delta t)}{P(T > t)} = \frac{F(t + \Delta t) - F(t)}{R(t)} = \frac{\Delta F(t)}{R(t)}$$

By dividing the probability by Δt and finding the limit as $\Delta t \to 0$ gives the hazard rate:

$$h(t) = \lim_{\Delta t \to 0} \frac{P(t < T < t + \Delta t | T > t)}{\Delta t} = \lim_{\Delta t \to 0} \frac{\Delta F(t)}{\Delta t R(t)} = \frac{f(t)}{R(t)}$$

The discrete hazard rate is defined as: (Xie, Gaudoin, et al. 2002)

$$h(k) = \frac{P(K = k)}{P(K \geq k)} = \frac{f(k)}{R(k - 1)}$$

This unintuitive result is due to a popular definition of $R^*(k) = \sum_{i=k}^{\infty} f(k)$ in which case $h(k) = f(k)/R^*(k)$. This definition has been avoided because it violates the formula $R(k) = 1 - F(k)$. The discrete hazard rate cannot be used in the same way as a continuous hazard rate with the following differences (Xie, Gaudoin, et al. 2002):

- $h(k)$ is defined as a probability and so is bounded by [0,1].
- $h(k)$ is not additive for series systems.
- For the cumulative hazard rate $H(k) = -\ln[R(k)] \neq \sum_{i=0}^{k} h(k)$
- When a set of data is analyzed using a discrete counterpart of the continuous distribution the values of the hazard rate do not converge.

A function called the second failure rate has been proposed (Gupta et al. 1997):

$$r(k) = \ln \frac{R(k - 1)}{R(k)} = -\ln[1 - h(k)]$$

This function overcomes the previously mentioned limitations of the discrete hazard rate function and maintains the monotonicity property. For more information, the reader is referred to (Xie, Gaudoin, et al. 2002).

Care should be taken not to confuse the hazard rate with the *Rate of Occurrence of Failures (ROCOF)*. ROCOF is the probability that a failure (not necessarily the first) occurs in a short time interval. Unlike the hazard rate, the ROCOF is the absolute rate at which system failures occur and is not conditional on survival to time t. ROCOF is using in measuring the change in the rate of failures for repairable systems.

1.2.10.   Cumulative Hazard

The *cumulative hazard*, denoted as $H(t)$ in the continuous case is the area under the hazard rate function. This function is useful to calculate average failure rates.

$$H(t) = \int_0^t h(u)du = -\ln[R(t)]$$
$$H(k) = -\ln[R(k)]$$

For a discussion on the discrete cumulative hazard rate see *hazard rate*.

1.2.11.   Characteristic Function

The characteristic function of a random variable completely defines its probability distribution. It can be used to derive properties of the distribution from transformations of the random variable. (Billingsley 1995)

The characteristic function is defined as the expected value of the function $\exp(i\omega x)$ where x is the random variable of the distribution with a cdf $F(x)$, ω is a parameter that can have any real value and $i = \sqrt{-1}$:

$$\varphi_X(\omega) = E\left[e^{i\omega x}\right]$$
$$= \int_{-\infty}^{\infty} e^{i\omega x} F(x)\ dx$$

A useful property of the characteristic function is the sum of independent random variables is the product of the random variables characteristic function. It is often easier to use the natural log of the characteristic function when conducting this operation.

$$\varphi_{X+Y}(\omega) = \varphi_X(\omega)\varphi_Y(\omega)$$

$$\ln[\varphi_{X+Y}(\omega)] = \ln[\varphi_X(\omega)]\ln[\varphi_Y(\omega)]$$

For example, the addition of two exponentially distributed random variables with the same λ gives the gamma distribution with $k = 2$:

$$X{\sim}Exp(\lambda), \quad Y{\sim}Exp(\lambda)$$
$$\varphi_X(\omega) = \frac{i\lambda}{\omega + i\lambda}, \quad \varphi_Y(\omega) = \frac{i\lambda}{\omega + i\lambda}$$

$$\varphi_{X+Y}(\omega) = \varphi_X(\omega)\varphi_Y(\omega)$$
$$= \frac{-\lambda^2}{(\omega + i\lambda)^2}$$

$$X + Y {\sim} Gamma(k = 1, \lambda)$$

This is the characteristic function of the gamma distribution with $k = 2$.

The moment generating function can be calculated from the characteristic function:
$$\varphi_X(-i\omega) = M_X(\omega)$$

The nth raw moment can be calculated by differentiating the characteristic function n times. For more information on moments see section 1.3.2.

$$E[X^n] = i^{-n}\varphi_X^{(n)}(0)$$
$$= i^{-n}\left[\frac{d^n}{d\omega^n}\varphi_X(\omega)\right]$$

1.2.12.   Joint Distributions

Joint distributions are multivariate distributions with, d random variables $(d > 1)$. An example of a bivariate distribution $(d = 2)$ may be the distribution of failure for a vehicle tire which with random variables time, T, and distance travelled, X. The dependence between these two variables can be quantified in terms of correlation and covariance. See Section 1.3.3 for more discussion. For more on properties of multivariate distributions see (Rencher 1997).

Joint distributions can be derived from the conditional distributions. For the bivariate case with random variables X and Y:

$$f(x,y) = f(y|x)f(x) = f(x|y)f(y)$$

For the more general case of multivariate distribution:

$$\begin{aligned} f(x) &= f(x_1|x_2, \dots, x_d)f(x_2, \dots, x_d) \\ &= f(x_1)\, f(x_2|x_1) \dots f(x_{n-1}|x_1, \dots, x_{n-2})\, f(x_n|x_1, \dots, x_n) \end{aligned}$$

If the random variables are independent, their joint distribution is simply the product of the marginal distributions:

$$f(x) = \prod_{i=1}^{d} f(x_i) \quad where\ x_i \perp x_j for\ i \neq j$$

A general multivariate cumulative probability function with n random variables $(T_1, T_2, \dots, T_n)$ is defined as:

$$F(t_1, t_2, \dots, t_n) = P(T_1 \leq t_1, T_2 \leq t_2, \dots, T_n \leq t_n)$$

The survivor function is given as:

$$R(t_1, t_2, \dots, t_n) = P(T_1 > t_1, T_2 > t_2, \dots, T_n > t_n)$$

Different from univariate distributions is the relationship between the CDF and the survivor function (Georges et al. 2001):

$$F(t_1, t_2, \dots, t_n) + R(t_1, t_2, \dots, t_n) \leq 1$$

If $F(t_1, t_2, \dots, t_n)$ is differentiable then the probability density function is given as:

$$f(t_1, t_2, \dots, t_n) = \frac{\partial^n F(t_1, t_2, \dots, t_n)}{\partial t_1 \partial t_2 \dots \partial t_n}$$

For a discussion on the multivariate hazard rate functions and the construction of joint distributions from marginal distributions see (Singpurwalla 2006).

1.2.13.   Marginal Distribution

The marginal distribution of a single random variable for continuous and discrete joint distribution can be expressed, respectively, as:

$$f(x_1) = \int_{x_d} \dots \int_{x_3} \int_{x_2} f(x) \, dx_2 dx_3 \dots dx_d$$

$$f(k_1) = \sum_{k_2} \sum_{k_3} \dots \sum_{k_n} f(k)$$

1.2.14.   Conditional Distribution

If the value is known for some random variables the conditional distribution of the remaining random variables for continuous and discrete distributions are:

$$f(x_1|x_2, \dots, x_d) = \frac{f(x)}{f(x_2, \dots, x_d)} = \frac{f(x)}{\int_{x_1} f(x) \, dx_1}$$

$$f(k_1|k_2, \dots, k_d) = \frac{f(k)}{f(k_2, \dots, k_d)} = \frac{f(k)}{\sum_{k_1} f(x)}$$

1.2.15.   Bathtub Distributions

Elementary texts on reliability introduce the hazard rate of a system as a bathtub curve. The bathtub curve has three regions, infant mortality (decreasing failure rate), useful life (constant failure rate) and wear out (increasing failure rate). Bathtub distributions have not been a popular choice for modeling life distributions when compared to exponential, Weibull and lognormal distributions. This is because bathtub distributions are generally more complex without closed form moments and more difficult to estimate parameters.

Sometimes more complex shapes are required than simple bathtub curves, as such generalizations and modifications to the bathtub curves has been studied. These include an increase in the failure rate followed by a bathtub curve and rollercoaster curves (decreasing followed by unimodal hazard rate). For further reading including applications see (Lai & Xie 2006).

1.2.16.   Truncated Distributions

Truncation arises when the existence of a potential observation would be unknown if it were to occur in a certain range. An example of truncation is when the existence of a defect is unknown due to the defect's amplitude being less than the inspection threshold. The number of flaws below the inspection threshold is unknown. This is not to be confused with censoring which occurs when there is a bound for observing events. An example of right censoring is when a test is time terminated and the failures of the surviving components are not observed, however we know how many components were censored. (Meeker & Escobar 1998, p.266)

A truncated distribution is the conditional distribution that results from restricting the domain of another probability distribution. The following general formulas apply to truncated distribution functions, where $f_0(x)$ and $F_0(x)$ are the pdf and cdf of the non-truncated distribution. For further reading specific to common distributions see   (Cohen 1991)

Probability Distribution Function:

$$f(x) = \begin{cases} \dfrac{f_o(x)}{F_0(b) - F_0(a)} & for \; x \in (a, b] \\ 0 & otherwise \end{cases}$$

Cumulative Distribution Function:

$$F(x) = \begin{cases} 0 & for \; x \leq a \\ \dfrac{\int_a^x f_0(t) \; dt}{F_0(b) - F_0(a)} & for \; x \in (a, b] \\ 1 & for \; x > b \end{cases}$$

1.2.17. Summary

Table 2: Summary of important reliability function relationships

	$f(t)$	$F(t)$	$R(t)$	$h(t)$	$H(t)$
$f(t) =$	---	$F'(t)$	$-R'(t)$	$h(t)\exp\left\{-\int_0^t h(x)dx\right\}$	$-\dfrac{d\{\exp[-H(t)]\}}{dt}$
$F(t) =$	$\int_0^t f(x)dx$	----	$1-R(t)$	$1-\exp\left\{-\int_0^t h(x)dx\right\}$	$1-\exp\{-H(t)\}$
$R(t) =$	$1-\int_0^t f(x)dx$	$1-F(t)$	----	$\exp\left\{-\int_0^t h(x)dx\right\}$	$\exp\{-H(t)\}$
$h(t) =$	$\dfrac{f(t)}{1-\int_0^t f(x)dx}$	$\dfrac{F'(t)}{1-F(t)}$	$\dfrac{R'(t)}{R(t)}$	----	$H'(t)$
$H(t) =$	$-\ln\int_t^\infty f(x)dx$	$\ln\left\{\dfrac{1}{1-F(x)}\right\}$	$-\ln\{R(x)\}$	$\int_0^t h(x)dx$	---
$u(t) =$	$\dfrac{\int_0^\infty xf(t+x)dx}{\int_t^\infty f(x)dx}$	$\dfrac{\int_t^\infty[1-F(x)]dx}{1-F(t)}$	$\dfrac{\int_t^\infty R(x)dx}{R(t)}$	$\dfrac{\int_t^\infty \exp\left\{-\int_o^t h(x)dx\right\}dx}{\exp\left\{-\int_o^t h(x)dx\right\}}$	$\dfrac{\int_t^\infty \exp\{-H(x)\}\,dx}{\exp\{-H(x)\}}$

1.3. Distribution Properties

1.3.1. Median / Mode

The *median* of a distribution, denoted as $t_{0.5}$ is when the cdf and reliability function are equal to 0.5.

$$t_{0.5} = F^{-1}(0.5) = R^{-1}(0.5)$$

The *mode* is the highest point of the pdf, t_m. This is the point where a failure has the highest probability. Random samples taken from this distribution (for example during a Monte Carlo simulation) would occur most often around the mode.

1.3.2. Moments of Distribution

The first moment of continuous and discrete distribution are given by:

$$\mu_n = \int_{-\infty}^{\infty} (x - c)^n f(x)\, dx, \qquad \mu_n = \sum_i (k_j - c)^n f(k)$$

When $c = 0$ the moments, μ'_n, are called the *raw moments*, described as moments about the origin. In respect to probability distributions the first two raw moments are important. μ'_0 always equals one, and μ'_1 is the distributions *mean* which is the expected value of the random variable for the distribution:

$$\mu'_0 = \int_{-\infty}^{\infty} f(x)\, dx = 1, \qquad \mu'_0 = \sum_i f(k_i) = 1$$

mean = $E[X] = \mu$:

$$\mu'_1 = \int_{-\infty}^{\infty} x f(x)\, dx, \qquad \mu'_1 = \sum_i k_i f(k_i)$$

Some important properties of the expected value $E[X]$ when transformations of the random variable occur are:

$$E[X + b] = \mu_X + b$$

$$E[X \pm Y] = \mu_X \pm \mu_Y$$

$$E[aX] = a\mu_X$$

$$E[XY] = \mu_X \mu_Y + Cov(X, Y)$$

When $c = \mu$ the moments, μ_n, are called the *central moments*, described as moments about the mean. In this book, the first five central moments are important. μ_0 is equal to $\mu'_0 = 1$. μ_2 is the *variance* which quantifies the amount the random variable deviates from the mean. μ_3 and μ_4 are used to calculate the skewness and kurtosis, respectively.

$$\mu_0 = \int_{-\infty}^{\infty} f(x)\, dx = 1, \qquad \mu_0 = \sum_i f(k_i) = 1$$

$$\mu_1 = \int_{-\infty}^{\infty} (x - \mu)f(x)\, dx = 0, \qquad \mu_1 = \sum_i (k_i - \mu)f(k_i) = 0$$

variance $= E[(X - E[X])^2] = E[X^2] - \{E[X]\}^2 = \sigma^2$:

$$\mu_2 = \int_{-\infty}^{\infty} (x - \mu)^2 f(x)\, dx, \qquad \mu_2 = \sum_i (k_i - \mu)^2 f(k_i)$$

Some important properties of the variance exist when transformations of the random variable occur are:

$$Var[X + b] = Var[X]$$

$$Var[X \pm Y] = \sigma_X^2 + \sigma_Y^2 \pm 2Cov(X, Y)$$

$$Var[aX] = a^2 \sigma_X^2$$

$$Var[XY] = (Var(X) + [E(X)]^2)(Var(Y) + [E(Y)]^2) - [Cov(X, Y) + E[X].E(Y)]^2 + Cov(X^2, Y^2)$$

The *skewness* is a measure of the asymmetry of the distribution.

$$\gamma_1 = \frac{\mu_3}{\mu_2^{3/2}}$$

The *kurtosis* is a measure of the whether the data is peaked or flat.

$$\gamma_2 = \frac{\mu_4}{\mu_2^2}$$

1.3.3. Covariance

Covariance is a measure of the dependence between random variables.

$$Cov(X, Y) = E[(X - \mu_X)(Y - \mu_Y)] = E[XY] - \mu_X \mu_Y$$

A normalized measure of covariance is *correlation*, ρ. The correlation has the limits $-1 \le \rho \le 1$. When $\rho = 1$ the random variables have a linear dependency (i.e, an increase in X will result in the same increase in Y). When $\rho = -1$ the random variables have a negative linear dependency (i.e, an increase in X will result in the same decrease in Y). The relationship between covariance and correlation is:

$$\rho_{X,Y} = Corr(X, Y) = \frac{Cov(X, Y)}{\sigma_X \sigma_Y}$$

If the two random variables are independent than the correlation is equal to zero, however the reverse is not always true. If the correlation is zero the random variables does not need to be independent. For derivations and more information the reader is referred to (Dekking et al. 2007, p.138).

1.4. Parameter Estimation

1.4.1. Probability Plotting Paper

Most plotting methods transform the data available into a straight line for a specific distribution. From a line of best fit the parameters of the distribution can be estimated. Most plotting paper plots the random variable (time or demands) against the pdf, cdf or hazard rate and transform the data points to a linear relationship by adjusting the scale of each axis. Probability plotting is done using the following steps (Nelson 1982, p.108):

1. Order the data such that $x_1 \leq x_2 \leq \cdots \leq x_i \leq \cdots \leq x_n$.

2. Assign a rank to each failure. For complete data this is simply the value i. Censored data is discussed after step 7.

3. Calculate the plotting position. The cdf may simply be calculated as i/n however this produces a biased result, instead the following non-parametric *Blom estimates*, are recommended as suitable for many cases by (Kimball 1960):

$$\hat{h}(t_i) = \frac{1}{(n - i + 0.625)(t_{i+1} - t_i)}$$

$$\hat{F}(t_i) = \frac{i - 0.375}{n + 0.25}$$

$$\hat{R}(t_i) = \frac{n - i + 0.625}{(n + 0.25)}$$

$$\hat{f}(t_i) = \frac{1}{(n + 0.25)(t_{i+1} - t_i)}$$

Other proposed estimators are:

$$\text{Naive: } \hat{F}(t_i) = \frac{i}{n}$$

$$\text{Median (approximate): } \hat{F}(t_i) = \frac{i - 0.3}{n + 0.4}$$

$$\text{Midpoint: } \hat{F}(t_i) = \frac{i - 0.5}{n}$$

$$\text{Mean : } \hat{F}(t_i) = \frac{i}{n + 1}$$

$$\text{Mode: } \hat{F}(t_i) = \frac{i - 1}{n - 1}$$

4. Plot points on probability paper. The choice of distribution should be from experience, or multiple distributions should be used to assess the best fit. Probability paper is available from http://www.weibull.com/GPaper/.

5. Assess the data and chosen distributions. If the data plots in straight line then the distribution may be a reasonable fit.

6. Draw a line of best fit. This is a subjective assessment which minimizes the deviation of the points from the chosen line.

7. Obtained the desired information. This may be the distribution parameters or estimates of reliability or hazard rate trends.

When multiple failure modes are observed but failures didn't occur by the time observation was ended, units are censored and only one failure mode should be plotted with the other failures being treated as censored. Two popular methods to treat censored data two methods are:

Rank Adjustment Method. (Manzini et al. 2009, p.140) Here the adjusted rank, j_{t_i} is calculated only for non-censored units (with i_{t_i} still being the rank for all ordered times). This adjusted rank is used for step 2 with the remaining steps unchanged:

$$j_{t_i} = j_{t_{i-1}} + \frac{(n+1) - j_{t_{i-1}}}{2 + n - i_{t_i}}$$

Kaplan Meier Estimator. Here the estimate for reliability is:

$$\hat{R}(t_i) = \prod_{t_j < t_i} \left(1 - \frac{d}{n - i + 1}\right)$$

Where d is the number of failures in rank j (for non-grouped data $d = 1$). From this estimate a cdf can be given as $\hat{F}(t_i) = 1 - \hat{R}(t_i)$. For a detailed derivation and properties of this estimator see (Andersen et al. 1996, p.255)

Probability plots are fast and not dependent on complex numerical methods and can be used without a detailed knowledge of statistics. It provides a visual representation of the data for which qualitative statements can be made. It can be useful in estimating initial values for numerical methods. Limitation of this technique is that it is not objective and two different people making the same plot will obtain different answers. It also does not provide confidence intervals. For more detail of probability plotting the reader is referred to (Nelson 1982, p.104) and (Meeker & Escobar 1998, p.122)

1.4.2. Total Time on Test Plots

Total time on Test (TTT) plots is a graph which provides a visual representation of the hazard rate trend, i.e. increasing, constant or decreasing. This assists in identifying the distribution from which the data may come from. To plot TTT (Rinne 2008, p.334):

1. Order the data such that $x_1 \leq x_2 \leq \cdots \leq x_i \leq \cdots \leq x_n$.

2. Calculate the TTT positions:

$$TTT_i = \sum_{j=1}^{i} (n - j + 1)(x_j - x_{j-1}) \, ; i = 1, 2, \dots, n$$

3. Calculate the normalized TTT positions:

$$TTT_i^* = \frac{TTT_i}{TTT_n}; i = 1,2,\dots,n$$

4. Plot the points $\left(\frac{i}{n}, TTT_i^*\right)$.

5. Analyze graph:

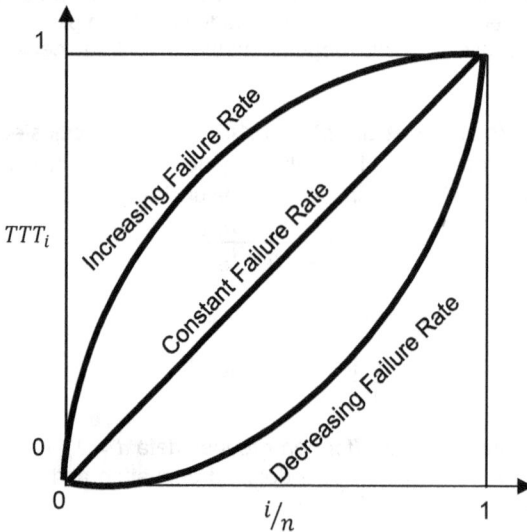

Figure 5: Time on test plot interpretation

Compared to probability plotting, TTT plots are simple, scale invariant and can represent any data set even those from different distributions on the same plot. However, it only provides an indication of failure rate properties and cannot be used directly to estimate parameters. For more information about TTT plots the reader is referred to (Rinne 2008, p.334).

1.4.3. Least Mean Square Regression

When the relationship between two variables, x and y is assumed linear ($y = mx + c$), an estimate of the line's parameters can be obtained from n sample data points, (x_i, y_i) using *least mean square (LMS) regression*. The least square method minimizes the square of the residual.

$$S = \sum_{i=1}^{n} r_i^2$$

The residual can be defined in many ways.

Minimize y residuals

$$r_i = y_i - f(x_i; m, c)$$

$$\hat{m} = \frac{n\sum x_i y_i - (\sum x_i)(\sum y_i)}{n\sum x_i^2 - (\sum x_i^2)^2}$$

$$\hat{c} = \frac{\sum y_i}{n} - \hat{m}\frac{\sum x_i}{n}$$

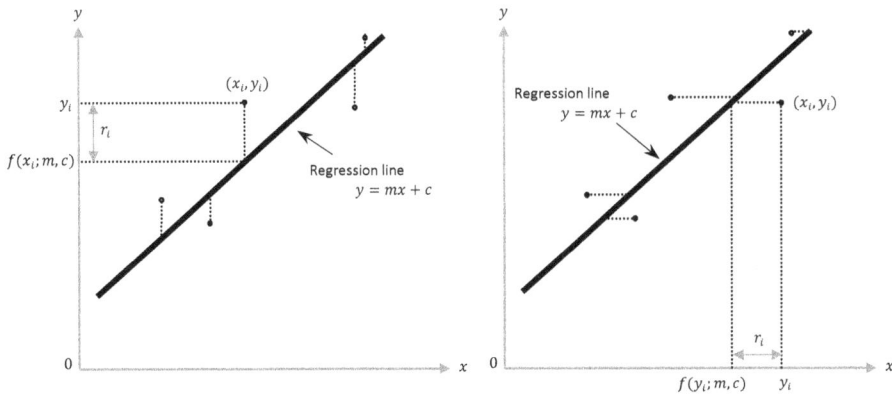

Minimize x residuals

$$r_i = x_i - f(y_i; m, c)$$

$$\hat{m} = \frac{n\sum y_i^2 - (\sum y_i^2)^2}{n\sum x_i y_i - (\sum x_i)(\sum y_i)}$$

$$\hat{c} = \frac{\sum y_i}{n} - \hat{m}\frac{\sum x_i}{n}$$

Figure 6: Left minimize y residual, right minimize x residual

The LMS method can be used to estimate the line of best fit when using plotting parameter estimation methods. When plotting on a regular scale in software such as Microsoft Excel, it is often easy to conduct linear least mean square (LMS) regression using in built functions. Where available this book provides the formulas to plot the sample data in a straight line in a regular scale plot. It also provides the transformation from the linear LMS regression estimates of $\hat{m}$ and $\hat{c}$ to the distribution parameter estimates.

For more on least square methods in a reliability engineering context see (Nelson 1990, p.167). MS regression can also be conducted on multivariate distributions, see (Rao & Toutenburg 1999) and can also be conducted on non-linear data directly, see (Björck 1996).

1.4.4. Method of Moments

To estimate the distribution parameters using the *method of moments* the sample moments are equated to the parameter moments and solved for the unknown parameters. The following sample moments can be used:

The sample mean is given as:

$$\bar{x} = \frac{1}{n}\sum_{i=1}^{n} x_i$$

The unbiased sample variance is given as:

$$S^2 = \frac{1}{n-1}\sum_{i=1}^{n}(x_i - \bar{x})^2$$

Method of moments is not as accurate as Bayesian or maximum likelihood estimates, but is easy and fast to calculate. The method of moment estimates is often used as a starting point for numerical methods to optimize maximum likelihood and least square estimators.

1.4.5. Maximum Likelihood Estimates

Maximum likelihood estimates (MLE) are based on a frequentist approach to parameter estimation usually obtained by maximizing the natural logarithm of the likelihood function.

$$\Lambda(\theta|E) = \ln\{L(\theta|E)\}$$

Algebraically this is done by solving the first order partial derivatives of the log-likelihood function. This calculation has been included in this book for distributions where the result is in closed form. Otherwise, the log-likelihood function can be maximized directly using computer-based numerical simulation methods.

MLE for $\hat{\theta}$ is obtained by solving for θ:

$$\frac{\partial\Lambda}{\partial\theta} = 0$$

Denote the true parameters of the distribution as θ_0, MLEs have the following properties (Rinne 2008, p.406):
- **Consistency.** As the number of samples increases the difference between the estimated and actual parameter decreases:
$$\operatorname*{plim}_{n\to\infty} \hat{\theta} = \theta$$

- **Asymptotic normality.**
$$\lim_{n\to\infty} \hat{\theta} \sim Norm(\theta_0, [I_n(\theta_0)]^{-1})$$

where $I_n(\theta) = nI(\theta)$ is the Fisher information matrix. Therefore $\hat{\theta}$ is asymptotically unbiased:
$$\lim_{n\to\infty} E[\hat{\theta}] = \theta_0$$

- **Asymptotic efficiency.**
$$\lim_{n\to\infty} Var[\hat{\theta}] = [I_n(\theta_0)]^{-1}$$

- **Invariance.** The MLE of $f(\theta_0)$ is $f(\hat{\theta})$ if $f(.)$ is a continuous and continuously differentiable function.

The advantages of MLE are that it is a very common technique that has been widely published and is implemented in many software packages. The MLE method can easily handle censored data. The disadvantage to MLE is the bias introduced for small sample sizes and unbounded estimates may result when no failures have been observed. The numerical optimization of the log-likelihood function may be non-trivial with high sensitivity to starting values and the presence of local maximums.

For more information in a reliability context see (Nelson 1990, p.284).

1.4.6. Bayesian Estimation

Bayesian estimation uses a subjective interpretation of the theory of probability and for parameter point estimation and confidence intervals uses Bayes' rule to update our state of knowledge of the unknown of interest (UIO). Recall from Section 1.1.5 Bayes rule,

$$\pi(\theta|E) = \frac{\pi_o(\theta)\, L(E|\theta)}{\int \pi_o(\theta)\, L(E|\theta)\, d\theta}, \qquad P(\theta|E) = \frac{P(\theta)\, P(E|\theta)}{\sum_i P(E|\theta_i) P(\theta)}$$

respectively for continuous and discrete forms of variable of θ.

The Prior Distribution $\pi_o(\theta)$

The prior distribution is probability distribution of the UOI, θ, which captures our state of knowledge of θ prior to the evidence being observed. It is common for this distribution to represent soft evidence or intervals about the possible values of θ. If the distribution is dispersed it represents little being known about the parameter. If the distribution is concentrated in an area then it reflects a good knowledge about the likely values of θ.

Prior distributions should be a proper probability distribution of θ. A distribution is proper when it integrates to one and improper otherwise. The prior should also not be selected based on the form of the likelihood function and the data used to build the likelihood function. When the prior has a constant, which does not affect the posterior distribution (such as improper priors) it will be omitted from the formulas within this book.

Non-informative Priors. Occasions arise when it is not possible to express a subjective prior distribution due to lack of adequate information, time or cost. Alternatively, a subjective prior distribution may introduce unwanted bias through model convenience (conjugates) or due to elicitation methods. In such cases a non-informative prior may be desirable. The following methods exist for creating a non-informative prior (Yang and Berger 1998):

- **Principle of Indifference - Improper Uniform Priors.** An equal probability is assigned across all the plausible values of the parameter. This is done using an improper uniform distribution with a constant, usually 1, over the range of the possible values for θ. When placed in Bayes formula the constant cancels out, however the denominator is integrated over all possible values of θ. In most cases this prior distribution will result in a proper posterior, but not always. Improper Uniform Priors may be chosen to enable the use of conjugate priors.

  For example, using the exponential likelihood model, with an improper uniform prior, 1, over the limits $[0, \infty)$ with evidence of n_F failures in total time, t_T:

  Prior: $\pi_0(\lambda) = 1 \propto Gamma(1,0)$

  Likelihood: $L(E|\lambda) = \lambda^{n_F} e^{-\lambda t_T}$

$$\text{Posterior: } \pi(\lambda|E) = \frac{1. L(E|\lambda)}{1. \int_0^\infty L(E|\lambda)\, d\lambda}$$

Using conjugate relationship (see *Conjugate* Priors for calculations):

$$\lambda \sim Gamma(\lambda; 1 + n_F, t_T)$$

- **Principle of Indifference - Proper Uniform Priors.** An equal probability is assigned across the values of the parameter within a range defined by the uniform distribution. The uniform distribution is obtained by estimating the far left and right bounds (a and b) of the parameter θ giving $\pi_o(\theta) = \frac{1}{b-a} = c$, where c is a constant. When placed in Bayes formula the constant cancels out, however the denominator is integrated over the bound $[a, b]$. Care needs to be taken in choosing a and b because no matter how much evidence suggests otherwise the posterior distribution will always be zero outside these bounds.

Using an exponential likelihood model, with a proper uniform prior, c, over the limits $[a, b]$ with evidence of n_F failures in total time, t_T:

$$\text{Prior: } \pi_0(\lambda) = \frac{1}{b-a} = c \propto Truncated\ Gamma(1,0)$$

$$\text{Likelihood: } L(E|\lambda) = \lambda^{n_F} e^{-\lambda t_T}$$

$$\text{Posterior: } \pi(\lambda|E) = \frac{c. L(E|\lambda)}{c. \int_a^b L(E|\lambda)\, d\lambda} \quad \text{for } a \leq \lambda \leq b$$

Using conjugate relationship this results in a truncated Gamma distribution:

$$\pi(\lambda) = \begin{cases} c.\ Gamma(\lambda; 1 + n_F, t_T) & \text{for } a \leq \lambda \leq b \\ 0 & \text{otherwise} \end{cases}$$

- **Jeffrey's Prior.** Proposed by Jeffery in 1961, this prior is defined as $\pi_0(\theta) = \sqrt{det(I_\theta)}$ where I_θ is the Fisher information matrix. This derivation is motivated by the fact that it is not dependent upon the set of parameter variables that is chosen to describe parameter space. Jeffery himself suggested the need to make ad hoc modifications to the prior to avoid problems in multidimensional distributions. Jeffery's prior is normally improper. (Bernardo et al. 1992)

- **Reference Prior.** Proposed by Bernardo in 1979, this prior maximizes the expected posterior information from the data, therefore reducing the effect of the prior. When there is no nuisance parameters and certain regularity conditions are satisfied the reference-prior is identical to the Jeffrey's prior. Due to the need to order or group the importance of parameters, it may occur that different posteriors will result from the same data depending on the importance the user places on each parameter. This prior overcomes the problems which arise when using Jeffery's prior in multivariate applications.

- **Maximal Data Information Prior (MDIP).** Developed by Zelluer in 1971 maximizes the likelihood function with relation to the prior. (Berry et al. 1995, p.182)

For further detail on the differences between each type of non-informative prior see (Berry et al. 1995, p.179)

Conjugate Priors. Calculating posterior distributions can be extremely complex and, in most cases, requires expensive computations. A special case exists however by which the posterior distribution is of the same form as the prior distribution. The Bayesian updating mathematics can be reduced to simple calculations to update the model parameters. As an example, the gamma function is a conjugate prior to a Poisson likelihood function:

$$\text{Prior: } \pi_o(\lambda) = \frac{\beta^\alpha \lambda^{\alpha-1}}{\Gamma(\alpha)} e^{-\beta\lambda}$$

$$\text{Likelihood: } L_i(t_i|\lambda) = \frac{\lambda_i^k t_i^k}{k_i!} e^{-\lambda t_i}$$

$$\text{Likelihood of evidence: } L(E|\lambda) = \prod_{i=1}^{n_F} L_i(t_i|\lambda) = \frac{\lambda^{\Sigma k} \prod t_i^k}{\prod k_i!} e^{-\lambda \Sigma t_i}$$

$$\text{Posterior: } \pi(\lambda|E) = \frac{\pi_o(\lambda) L(E|\lambda)}{\int_0^\infty \pi_o(\lambda) L(E|\lambda) \, d\lambda}$$

$$= \frac{\frac{\beta^\alpha \lambda^{\alpha-1} \lambda^{\Sigma k} \prod t_i^k}{\Gamma(\alpha)\prod k_i!} e^{-\beta\lambda} e^{-\lambda\Sigma t_i}}{\int_0^\infty \frac{\beta^\alpha \lambda^{\alpha-1} \lambda^{\Sigma k} \prod t_i^k}{\Gamma(\alpha)\prod k_i!} e^{-\beta\lambda} e^{-\lambda\Sigma t_i} \, d\lambda}$$

$$= \frac{\lambda^{\alpha-1+\Sigma k} e^{-\lambda(\beta+\Sigma t_i)}}{\int_0^\infty \lambda^{\alpha-1+\Sigma k} e^{-\lambda(\beta+\Sigma t_i)} \, d\lambda}$$

Using the identity $\Gamma(z) = \int_o^\infty x^{z-1} e^{-x} \, dx$ we can calculate the denominator using the change of variable $u = \lambda(\beta + \Sigma t_i)$. This results in $\lambda = \frac{u}{\beta+\Sigma t_i}$ and $d\lambda = \frac{du}{\beta+\Sigma t_i}$, with the limits of u the same as λ. Substituting back into the posterior equation gives:

$$\pi(\lambda|E) = \frac{\lambda^{\alpha-1+\Sigma k} e^{-\lambda(\beta+\Sigma t_i)}}{\frac{1}{\beta+\Sigma t_i} \int_0^\infty \left(\frac{u}{\beta+\Sigma t_i}\right)^{\alpha-1+\Sigma k} e^{-u} \, du}$$

$$= \frac{\lambda^{\alpha-1+\Sigma k} e^{-\lambda(\beta+\Sigma t_i)}}{\frac{1}{(\beta+\Sigma t_i)^{\alpha+\Sigma k}} \int_0^\infty u^{\alpha-1+\Sigma k} e^{-u} \, du}$$

Let $z = \alpha + \sum k$

$$\pi(\lambda|E) = \frac{\lambda^{\alpha-1+\sum k} e^{-\lambda(\beta+\sum t_i)}}{\frac{1}{(\beta+\sum t_i)^{\alpha+\sum k}} \int_0^\infty u^{z-1} e^{-u} \, du}$$

Using $\Gamma(z) = \int_0^\infty x^{z-1} e^{-x} \, dx$:

$$\pi(\lambda|E) = \frac{\lambda^{\alpha-1+\sum k} (\beta+\sum t_i)^{\alpha+\sum k}}{\Gamma(\alpha+\sum k)} e^{-\lambda(\beta+\sum t_i)}$$

Let $\alpha' = \alpha + \sum k$, $\beta' = \beta + \sum t_i$:

$$\pi(\lambda|E) = \frac{\lambda^{\alpha'-1}\beta'^{\alpha'}}{\Gamma(\alpha')} e^{-\beta'\lambda}$$

As can be seen the posterior is a gamma distribution with the parameters $\alpha' = \alpha + \sum k$, $\beta' = \beta + \sum t_i$. Therefore, the prior and posterior are of the same form, and Bayes' rule does not need to be re-calculated for each update. Instead the user can simply update the parameters with the new evidence.

The Likelihood Function $L(E|\theta)$

The reader is referred to Section 1.1.6 for a discussion on the construction of the likelihood function.

The Posterior Distribution $\pi(\theta|E)$

The posterior distribution is a probability distribution of the UOI, θ, which captures our state of knowledge of θ including all prior information and the evidence.

Point Estimate. From the posterior distribution we may want to give an estimate of θ. The Bayesian estimator when using a quadratic loss function is the posterior mean (Christensen & Huffman 1985):

$$E[\pi(\theta|E)] = \int \theta\pi(\theta|E) \, d\theta = \mu_\pi$$

For more information on utility, loss functions and estimators in a Bayesian context see (Berger 1993).

1.4.7. Confidence Intervals

Assuming a random variable is distributed by a given distribution, there exists the true distribution parameters, $\boldsymbol{\theta_0}$, which is unknown. The parameter, mean, may or may not be close to the true parameter values. Confidence intervals provide the range over which the true parameter values may exist with a certain level of confidence. Confidence intervals only quantify uncertainty due to sampling error arising from a limited number of samples. Uncertainty due to incorrect model selection or incorrect assumptions is not included. (Meeker & Escobar 1998, p.49)

Increasing the desired confidence γ results in an increased confidence interval. Increasing the sample size generally decreases the confidence interval. There are many methods to calculate confidence intervals. Some popular methods are:

- **Exact Confidence Intervals.** It may be mathematically shown that the parameter of a distribution itself follows a distribution. In such cases exact confidence intervals can be derived. This is only the case in very few distributions.

- **Fisher Information Matrix** (Nelson 1990, p.292). For a large number of samples, the asymptotic normal property can be used to estimate confidence intervals:

$$\lim_{n\to\infty} \hat{\theta} \sim Norm(\theta_0, [nI(\theta_0)]^{-1})$$

Combining this with the asymptotic property $\hat{\theta} \to \theta_0$ as $n \to \infty$ gives the following estimate for the distribution of $\hat{\theta}$:

$$\lim_{n\to\infty} \hat{\theta} \sim Norm\left(\hat{\theta}, [J_n(\hat{\theta})]^{-1}\right)$$

$100\gamma\%$ approximate confidence intervals are calculated using percentiles of the normal distribution. If the range of θ is unbounded $(-\infty, \infty)$ the approximate two sided confidence intervals are:

$$\underline{\theta_\gamma} = \hat{\theta} - \Phi^{-1}\left(\frac{1+\gamma}{2}\right)\sqrt{[J_n(\hat{\theta})]^{-1}}$$

$$\overline{\theta_\gamma} = \hat{\theta} + \Phi^{-1}\left(\frac{1+\gamma}{2}\right)\sqrt{[J_n(\hat{\theta})]^{-1}}$$

If the range of θ is $(0, \infty)$ the approximate two-sided confidence intervals are:

$$\underline{\theta_\gamma} = \hat{\theta}.\exp\left[\frac{\Phi^{-1}\left(\frac{1+\gamma}{2}\right)\sqrt{[J_n(\hat{\theta})]^{-1}}}{-\hat{\theta}}\right]$$

$$\overline{\theta_\gamma} = \hat{\theta}.\exp\left[\frac{\Phi^{-1}\left(\frac{1+\gamma}{2}\right)\sqrt{[J_n(\hat{\theta})]^{-1}}}{\hat{\theta}}\right]$$

If the range of θ is $(0,1)$ the approximate two sided confidence intervals are:

$$\underline{\theta_\gamma} = \hat{\theta}.\left\{\hat{\theta} + (1-\hat{\theta})\exp\left[\frac{\Phi^{-1}\left(\frac{1+\gamma}{2}\right)\sqrt{[J_n(\hat{\theta})]^{-1}}}{\hat{\theta}(1-\hat{\theta})}\right]\right\}^{-1}$$

$$\overline{\theta_\gamma} = \hat{\theta}.\left\{\hat{\theta} + (1-\hat{\theta})\exp\left[\frac{\Phi^{-1}\left(\frac{1+\gamma}{2}\right)\sqrt{[J_n(\hat{\theta})]^{-1}}}{-\hat{\theta}(1-\hat{\theta})}\right]\right\}^{-1}$$

The advantage of this method is it can be calculated for all distributions and is easy to calculate. The disadvantage is that the assumption of a normal distribution is asymptotic and so sufficient data is required for the confidence interval estimate to be accurate. The number of samples needed for an accurate estimate changes from distribution to distribution. It also produces symmetrical confidence intervals which may be very inaccurate. For more information see (Nelson 1990, p.292).

- **Likelihood Ratio Intervals** (Nelson 1990, p.292). The test statistic for the likelihood ratio is:

$$D = 2[\Lambda(\hat{\theta}) - \Lambda(\theta)]$$

D is approximately Chi-Square distributed with one degree of freedom.

$$D = 2[\Lambda(\hat{\theta}) - \Lambda(\theta)] \le \chi^2(\gamma; 1)$$

Where γ is the $100\gamma\%$ confidence interval for θ. The two-sided confidence limits $\underline{\theta_\gamma}$ and $\overline{\theta_\gamma}$ are calculated by solving:

$$\Lambda(\theta) = \Lambda(\hat{\theta}) - \frac{\chi^2(\gamma; 1)}{2}$$

The limits are normally solved numerically. The likelihood ratio intervals are always within the limits of the parameter and gives asymmetrical confidence limits. It is much more accurate than the Fisher information matrix method particularly for one-sided limits although it is more complicated to calculate. This method must be solved numerically and so will not be discussed further in this book.

- **Bayesian Confidence Intervals.** In Bayesian statistics the uncertainty of a parameter, θ, is quantified as a distribution $\pi(\theta)$. Therefore, the two sided $100\gamma\%$ confidence intervals are found by solving:

$$\frac{1 - \gamma}{2} = \int_{-\infty}^{\theta_\gamma} \pi(\theta)\,d\theta, \qquad \frac{1 + \gamma}{2} = \int_{\theta_\gamma}^{\infty} \pi(\theta)\,d\theta$$

Other methods exist to calculate approximate confidence intervals. A summary of some techniques used in reliability engineering is included in (Lawless 2002).

1.5. Related Distributions

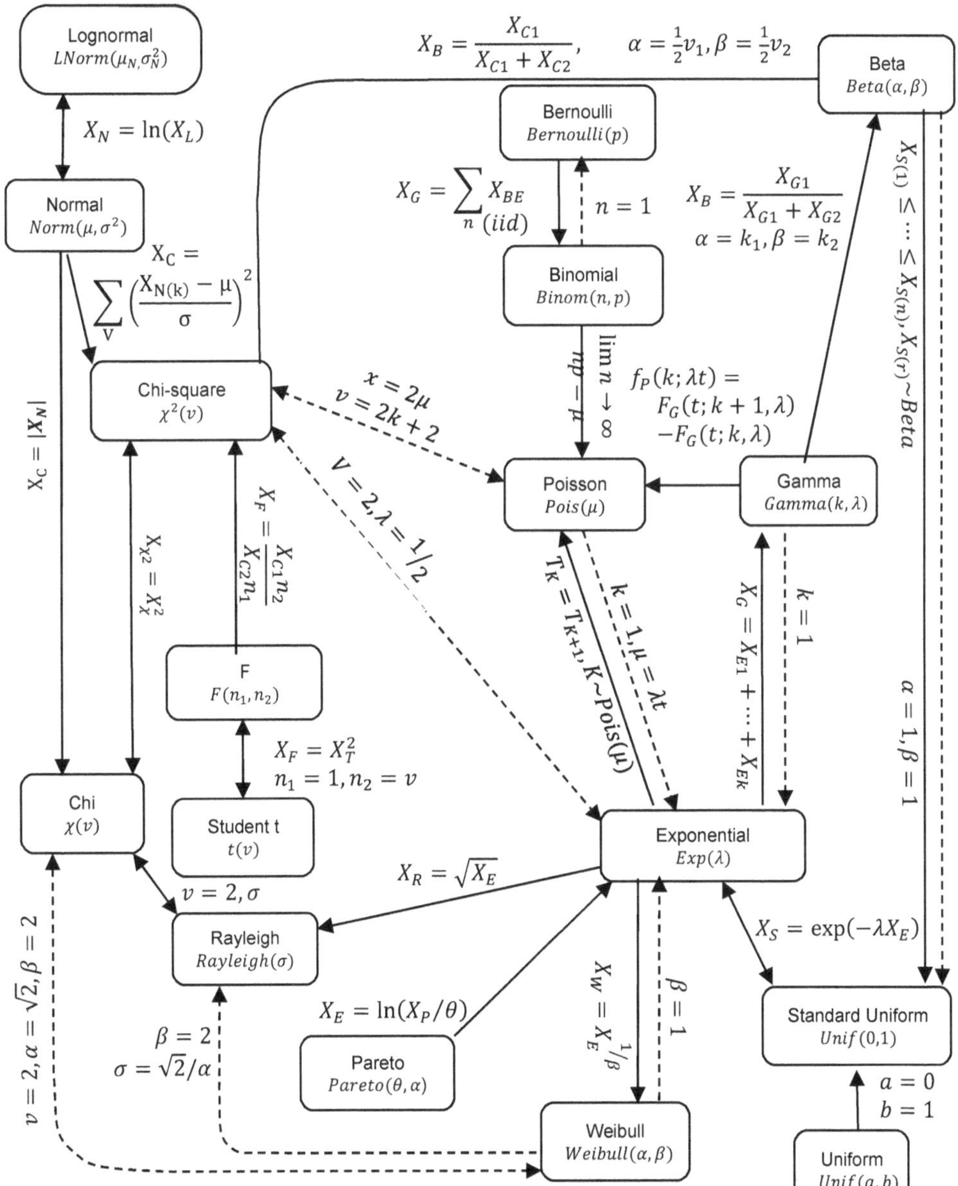

Figure 7: Relationships between common distributions (Leemis & McQueston 2008).

Many relations are not included such as central limit convergence to the normal distribution and many transforms which would have made the figure unreadable. For further details refer to individual sections and (Leemis & McQueston 2008).

1.6. Supporting Functions

1.6.1. Beta Function $B(x, y)$

$B(x, y)$ is the Beta function and is the Euler integral of the first kind.

$$B(x,y) = \int_0^1 u^{x-1}(1-u)^{y-1}du$$

Where $x > 0$ and $y > 0$.

Relationships:

$$B(x,y) = B(y,x)$$
$$B(x,y) = \frac{\Gamma(x)\Gamma(y)}{\Gamma(x+y)}$$
$$B(x,y) = \sum_{n=0}^{\infty} \frac{\binom{n-y}{n}}{x+n}$$

More formulas, definitions and special values can be found in the Digital Library of Mathematical Functions on the National Institute of Standards and Technology (NIST) website, http://dlmf.nist.gov.

1.6.2. Incomplete Beta Function $B_t(t; x, y)$

$B_t(t; x, y)$ is the incomplete Beta function expressed by:

$$B_t(t;x,y) = \int_0^t u^{x-1}(1-u)^{y-1}du$$

1.6.3. Regularized Incomplete Beta Function $I_t(t; x, y)$

$I_t(t|x, y)$ is the regularized incomplete Beta function:

$$I_t(t|x,y) = \frac{B_t(t;x,y)}{B(x,y)}$$

$$= \sum_{j=x}^{x+y-1} \frac{(x+y-1)!}{j!(x+y-1-j)!} \cdot t^j(1-t)^{x+y-1-j}$$

Properties:

$$I_0(0;\ x,y) = 0$$

$$I_1(1;\ x,y) = 1$$

$$I_t(t;\ x,y) = 1 - I(1-t;\ y,x)$$

1.6.4. Complete Gamma Function $\Gamma(k)$

$\Gamma(k)$ is a generalization of the factorial function $k!$ to include non-integer values.

For $k > 0$

$$\Gamma(k) = \int_0^\infty t^{k-1} e^{-t} dt$$

$$= \left[-t^{k-1} e^{-t} \right]_0^\infty + (k-1) \int_0^\infty t^{k-2} e^{-t} dt$$

$$= (k-1) \int_0^\infty t^{k-2} e^{-t} dt$$

$$= (k-1)\Gamma(k-1)$$

When k is an integer:

$$\Gamma(k) = (k-1)!$$

Special values:

$$\Gamma(1) = 1$$

$$\Gamma(2) = 1$$

$$\Gamma\left(\frac{1}{2}\right) = \sqrt{\pi}$$

Relation to the incomplete gamma functions:

$$\Gamma(k) = \Gamma(k,t) + \gamma(k,t)$$

More formulas, definitions and special values can be found in the Digital Library of Mathematical Functions on the National Institute of Standards and Technology (NIST) website, http://dlmf.nist.gov.

1.6.5. Upper Incomplete Gamma Function $\Gamma(k, t)$

For $k > 0$

$$\Gamma(k,t) = \int_t^\infty x^{k-1} e^{-x} dx$$

When k is an integer:

$$\Gamma(k,t) = (k-1)! \, e^{-t} \sum_{n=0}^{k-1} \frac{t^n}{n!}$$

More formulas, definitions and special values can be found on the NIST website, http://dlmf.nist.gov.

1.6.6. Lower Incomplete Gamma Function $\gamma(k, t)$

For $k > 0$

$$\gamma(k,t) = \int_0^t x^{k-1} e^{-x} dx$$

When k is an integer:

$$\gamma(k,t) = (k-1)! \left[1 - e^{-t} \sum_{n=0}^{k-1} \frac{t^n}{n!} \right]$$

More formulas, definitions and special values can be found on the NIST website, http://dlmf.nist.gov.

1.6.7. Digamma Function $\psi(x)$

$\psi(x)$ is the digamma function defined as:

$$\psi(x) = \frac{d}{dx}\ln[\Gamma(x)] = \frac{\Gamma'(x)}{\Gamma(x)} \quad for \; x > 0$$

1.6.8. Trigamma Function $\psi'(x)$

$\psi'(x)$ is the trigamma function defined as:

$$\psi'(x) = \frac{d^2}{dx^2}\ln\Gamma(x) = \sum_{i=0}^{\infty}(x+i)^{-2}$$

1.7. Referred Distributions

1.7.1. Inverse Gamma Distribution $IG(\alpha, \beta)$

The pdf to the inverse gamma distribution is:

$$f(x; \alpha, \beta) = \frac{\beta^\alpha}{\Gamma(\alpha)x^{\alpha+1}} \cdot e^{\frac{-\beta}{x}} \cdot I_x(0, \infty)$$

With mean:

$$\mu = \frac{\beta}{\alpha - 1} \text{ for } \alpha > 1$$

1.7.2. Student T Distribution $T(\alpha, \mu, \sigma^2)$

The pdf to the standard student t distribution with $\mu = 0$, $\sigma^2 = 1$ is:

$$f(x; \alpha) = \frac{\Gamma[(\alpha + 1)/2]}{\sqrt{\alpha\pi}\Gamma(\alpha/2)} \cdot \left(1 + \frac{x^2}{\alpha}\right)^{-\frac{\alpha+1}{2}}$$

The generalized student t distribution is:

$$f(x; \alpha, \mu, \sigma^2) = \frac{\Gamma[(\alpha + 1)/2]}{\sigma\sqrt{\alpha\pi}\Gamma(\alpha/2)} \cdot \left(1 + \frac{(x - \mu)^2}{\alpha\sigma^2}\right)^{-\frac{\alpha+1}{2}}$$

With mean

$$\mu = \mu$$

1.7.3. F Distribution $F(n_1, n_2)$

Also known as the Variance Ratio or Fisher-Snedecor distribution the pdf is:

$$f(x; \alpha) = \frac{1}{xB\left(\frac{n_1}{2}, \frac{n_2}{2}\right)} \cdot \sqrt{\frac{(n_1 x)^{n_1} \cdot n_2^{n_2}}{(n_1 x + n_2)^{\{n_1 + n_2\}}}}$$

With cdf:

$$I_t\left(\frac{n_1}{2}, \frac{n_2}{2}\right), \qquad where \ t = \frac{n_1 x}{n_1 x + n_2}$$

1.7.4. Chi-Square Distribution $\chi^2(v)$

The pdf to the chi-square distribution is:

$$f(x; v) = \frac{x^{(v-2)/2} \exp\left\{-\frac{x}{2}\right\}}{2^{v/2}\Gamma\left(\frac{v}{2}\right)}$$

With mean:

$$\mu = v$$

1.7.5. Hypergeometric Distribution $HyperGeom(k; n, m, N)$

The hypergeometric distribution calculates probability of k successes in n Bernoulli trials from population N containing m success without replacement. $p = m/N$. The pdf to the hypergeometric distribution is:

$$f(k; n, m, N) = \frac{\binom{m}{k}\binom{N-m}{n-k}}{\binom{N}{n}}$$

With mean:

$$\mu = \frac{nm}{N}$$

1.7.6. Wishart Distribution $Wishart_d(x; \Sigma, n)$

The Wishart distribution is the multivariate generalization of the gamma distribution. The pdf is given as:

$$f_d(x; \Sigma, n) = \frac{|x|^{\frac{1}{2}(n-d-1)}}{2^{nd/2}|\Sigma|^{n/2}\Gamma_d\left(\frac{n}{2}\right)} \exp\left\{-\frac{1}{2}tr\left(x^{-1}\Sigma\right)\right\}$$

With mean:

$$\mu = n\Sigma$$

1.8. Nomenclature and Notation

Functions are presented in the following form:
$$f(random\ variables\ ;\ parameters\ |given\ values)$$

n	In continuous distributions the number of items under test $= n_f + n_s + n_I$. In discrete distributions the total number of trials.
n_F	The number of items which failed before the conclusion of the test.
n_S	The number of items which survived to the end of the test.
n_I	The number of items which have interval data
t_i^F, t_i	The time at which a component fails.
t_i^S	The time at which a component survived to. The item may have been removed from the test for a reason other than failure.
t_i^{UI}	The upper limit of a censored interval in which an item failed
t_i^{LI}	The lower limit of a censored interval in which an item failed
t_L	The lower truncated limit of sample.
t_U	The upper truncated limit of sample.
t_T	Time on test $= \sum t_i + \sum t_s$
X or T	Continuous random variable (T is normally a random time)
K	Discrete random variable
x or t	A continuous random variable with a known value
k	A discrete random variable with a known value
$\hat{x}$	The *hat* denotes an estimated value
$\mathbf{x}$	A bold symbol denotes a vector or matrix
θ	Generalized unknown of interest (UOI)
$\overline{\theta}$	Upper confidence interval for UOI
$\underline{\theta}$	Lower confidence interval for UOI
$X \sim Norm_d$	The random variable X is distributed as a d-variate normal distribution.

2. Common Life Distributions

2.1. Exponential Continuous Distribution

Probability Density Function - f(t)

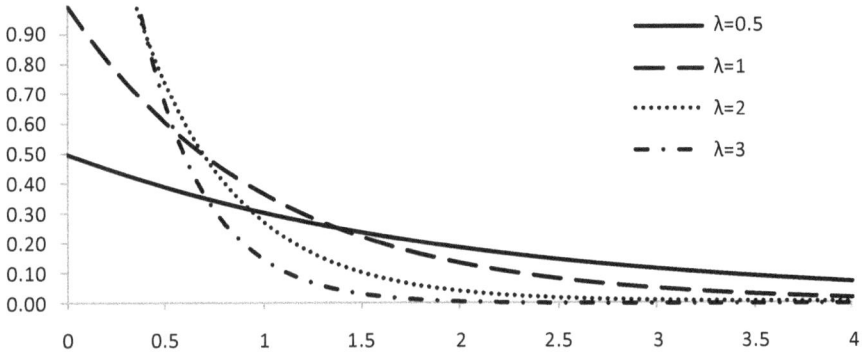

Cumulative Density Function - F(t)

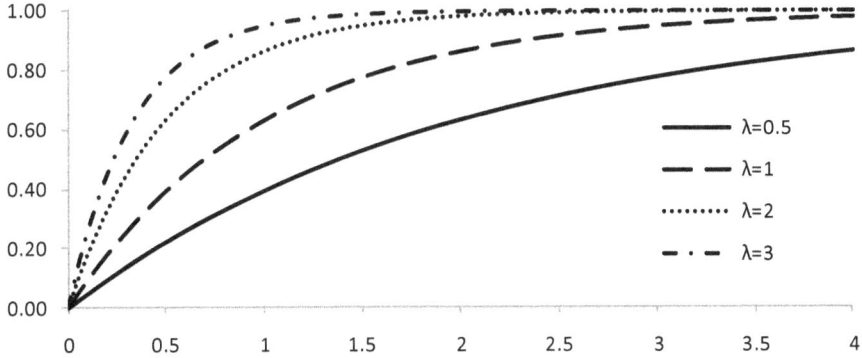

Hazard Rate - h(t)

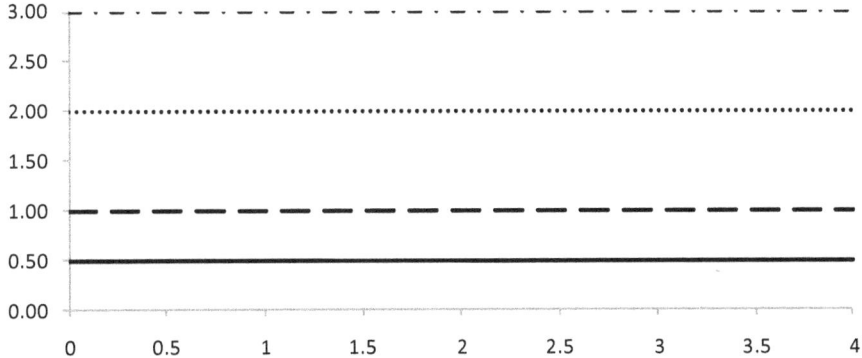

Parameters & Description

Parameters	λ	$\lambda > 0$	Scale Parameter: Equal to the hazard rate.
Limits		$t \geq 0$	

Function	Time Domain	Laplace Domain
PDF	$f(t) = \lambda e^{-\lambda t}$	$f(s) = \dfrac{\lambda}{\lambda + s}$, $s > -\lambda$
CDF	$F(t) = 1 - e^{-\lambda t}$	$F(s) = \dfrac{\lambda}{s(\lambda + s)}$
Reliability	$R(t) = e^{-\lambda t}$	$R(s) = \dfrac{1}{\lambda + s}$
Conditional Survivor Function $P(T > x + t \mid T > t)$	$m(x) = e^{-\lambda x}$	$m(s) = \dfrac{1}{\lambda + s}$

t is the given time we know the component has survived to.
x is value of the random variable X representing the time after t.
Note: $x = 0$ at t.

Mean Residual Life	$u(t) = \dfrac{1}{\lambda}$	$u(s) = \dfrac{1}{\lambda s}$
Hazard Rate	$h(t) = \lambda$	$h(s) = \dfrac{\lambda}{s}$
Cumulative Hazard Rate	$H(t) = \lambda t$	$H(s) = \dfrac{\lambda}{s^2}$

Properties and Moments

Median	$\dfrac{ln(2)}{\lambda}$
Mode	0
Mean - 1st Raw Moment	$\dfrac{1}{\lambda}$
Variance - 2nd Central Moment	$\dfrac{1}{\lambda^2}$
Skewness - 3rd Central Moment	2
Excess kurtosis - 4th Central Moment	6
Characteristic Function	$\dfrac{i\lambda}{t + i\lambda}$
100α% Percentile Function	$t_\alpha = -\dfrac{1}{\lambda} \ln(1 - \alpha)$

Parameter Estimation

Plotting Method

	X-Axis	Y-Axis	
Least Mean Square - $y = mx + c$	t_i	$ln[1 - F(t_i)]$	$\hat{\lambda} = -m$

Likelihood Function

$$L(E|\lambda) = \lambda^{n_F} \underbrace{\prod_{i=1}^{n_F} e^{-\lambda.t_i^F}}_{\text{failures}} \cdot \underbrace{\prod_{i=1}^{n_S} e^{-t_i^S}}_{\text{survivors}} \cdot \underbrace{\prod_{i=1}^{n_I} \left(e^{-\lambda t_i^{LI}} - e^{-\lambda t_i^{UI}}\right)}_{\text{interval failures}}$$

Likelihood Functions

When there is no interval data this reduces to:

$$L(E|\lambda) = \lambda^{n_F} e^{-\lambda t_T} \quad where \quad t_T = \sum t_i^F + \sum t_i^S = total\ time\ on\ test$$

$$\Lambda(E|\lambda) = n_F.\ln(\lambda) - \underbrace{\sum_{i=1}^{n_F} \lambda t_i^F}_{\text{failures}} - \underbrace{\sum_{s=1}^{n_S} \lambda t_s}_{\text{survivors}} + \underbrace{\sum_{j=1}^{n_I} \ln\left(e^{-\lambda t_j^{LI}} - e^{-\lambda t_j^{RI}}\right)}_{\text{interval failures}}$$

Log-Likelihood Functions

When there is no interval data this reduces to:

$$\Lambda(E|\lambda) = n_F.\ln(\lambda) - \lambda t_T \quad where \quad t_T = \sum_{n_F} t_i^F + \sum_{n_S} t_j^S$$

solve for λ to get $\hat{\lambda}$:

$$\frac{\partial \Lambda}{\partial \lambda} = 0 \qquad \underbrace{\frac{n_F}{\lambda} - \sum_{i=1}^{n_F} t_i^F}_{\text{failures}} - \underbrace{\sum_{s=1}^{n_S} t_S^S}_{\text{survivors}} - \underbrace{\sum_{i=1}^{n_I} \left(\frac{t_i^{LI} e^{\lambda t_i^{LI}} - t_i^{RI} e^{\lambda t_i^{RI}}}{e^{\lambda t_i^{LI}} - e^{\lambda t_i^{RI}}}\right)}_{\text{interval failures}} = 0$$

Point Estimates

When there is only complete and right-censored data:

$$\hat{\lambda} = \frac{n_F}{t_T} \qquad where \quad t_T = \sum t_i^F + \sum t_i^S = total\ time\ in\ test$$

Fisher Information

$$I(\lambda) = \frac{1}{\lambda}$$

100γ% Confidence Interval

(excluding interval data)

	λ_{lower} - 2-Sided	λ_{upper} - 2-Sided	λ_{upper} - 1-Sided
Type I (Time Terminated)	$\dfrac{\chi^2_{\left(\frac{1-\gamma}{2}\right)}(2n_F)}{2t_T}$	$\dfrac{\chi^2_{\left(\frac{1+\gamma}{2}\right)}(2n_F + 2)}{2t_T}$	$\dfrac{\chi^2_{(\gamma)}(2n_F + 2)}{2t_T}$
Type II (Failure Terminated)	$\dfrac{\chi^2_{\left(\frac{1-\gamma}{2}\right)}(2n_F)}{2t_T}$	$\dfrac{\chi^2_{\left(\frac{1+\gamma}{2}\right)}(2n_F)}{2t_T}$	$\dfrac{\chi^2_{(\gamma)}(2n_F)}{2t_T}$

$\chi^2_{(\alpha)}$ is the α percentile of the Chi-squared distribution. (Modarres et al. 1999, pp.151-152) Note: These confidence intervals are only valid for complete and right-censored data or when approximations of interval data are used (such as the median). They are exact confidence bounds and therefore approximate methods such as use of the Fisher information matrix need not be used.

Bayesian

Non-informative Priors $\pi(\lambda)$
(Yang and Berger 1998, p.6)

Type	Prior	Posterior
Uniform Proper Prior with limits $\lambda \in [a,b]$	$\dfrac{1}{b-a}$	Truncated Gamma Distribution For $a \le \lambda \le b$ $c.\, Gamma(\lambda; 1 + n_F, t_T)$ Otherwise $\pi(\lambda) = 0$
Uniform Improper Prior with limits $\lambda \in [0, \infty)$	$1 \propto Gamma(1,0)$	$Gamma(\lambda; 1 + n_F, t_T)$
Jeffrey's Prior	$\dfrac{1}{\sqrt{\lambda}} \propto Gamma(\frac{1}{2}, 0)$	$Gamma(\lambda; \frac{1}{2} + n_F, t_T)$ when $\lambda \in [0, \infty)$
Novick and Hall	$\dfrac{1}{\lambda} \propto Gamma(0,0)$	$Gamma(\lambda; n_F, t_T)$ when $\lambda \in [0, \infty)$

where $t_T = \sum t_i^F + \sum t_j^S$ = total time in test

Conjugate Priors

UOI	Likelihood Model	Evidence	Dist. of UOI	Prior Para	Posterior Parameters
λ from $Exp(t; \lambda)$	Exponential	n_F failures in t_T unit of time	Gamma	k_0, Λ_0	$k = k_o + n_F$ $\Lambda = \Lambda_o + t_T$

Description, Limitations and Uses

Example

Three vehicle tires were run on a test area for 1000km have punctures at the following distances:
Tire 1: No punctures
Tire 2: 400km, 900km
Tire 3: 200km

Punctures are a random failure with constant failure rate therefore an exponential distribution would be appropriate. Due to an exponential distribution being homogeneous in time, the renewal process of the second tire failing twice with a repair can be considered as two separate tires on test with single failures. See the example in Section 1.1.6.

Total distance on test is $3 \times 1000 = 3000$km. Total number of failures is 3. Therefore, using MLE the estimate of λ:

$$\hat{\lambda} = \frac{n_F}{t_T} = \frac{3}{3000} = 1\text{E-3}$$

With 90% confidence interval (distance terminated test):

$$\left[\frac{\chi^2_{(0.05)}(6)}{6000} = 0.272E\text{-}3, \qquad \frac{\chi^2_{(0.95)}(8)}{6000} = 2.584E\text{-}3 \right]$$

A Bayesian point estimate using the Jeffery non-informative improper prior $Gamma(\frac{1}{2}, 0)$, with posterior $Gamma(\lambda; 3.5, 3000)$ has a point estimate:

$$\hat{\lambda} = E[Gamma(\lambda; 3.5, 3000)] = \frac{3.5}{3000} = 1.1\dot{6}E - 3$$

with 90% confidence two-sided interval using inverse Gamma cdf:
$$[F_G^{-1}(0.05) = 0.361E\text{-}3, \qquad F_G^{-1}(0.95) = 2.344E\text{-}3]$$

Characteristics

Constant Failure Rate. The exponential distribution is defined by a constant failure rate, λ. This means the component is not subject to wear or accumulation of damage as time increases.

$f(0) = \lambda$. As can be seen, λ is the initial value of the distribution. Increases in λ increase the probability density at $f(0)$.

HPP. The exponential distribution is the time to failure distribution of a single event in the Homogeneous Poisson Process (HPP).

$$T \sim Exp(t; \lambda)$$

Scaling property

$$aT \sim Exp\left(t; \frac{\lambda}{a}\right)$$

Minimum property

$$\min\{T_1, T_2, \ldots, T_n\} \sim Exp\left(t; \sum_{i=1}^{n} \lambda_i\right)$$

Variate Generation property

$$F^{-1}(u) = \frac{\ln(1-u)}{-\lambda}, \quad 0 < u < 1$$

Memoryless property.

$$\Pr(T > t + x | T > t) = \Pr(T > x)$$

Properties from (Leemis & McQueston 2008).

Applications

No Wear out. The exponential distribution is used to model occasions when there is no wear out or cumulative damage. It can be used to approximate the failure rate in a component's useful life period (after burn in and before wear out).

Homogeneous Poisson Process (HPP). The exponential distribution is used to model the inter arrival times in a repairable system or the arrival times in queuing models. See Poisson and Gamma distribution for more detail.

Electronic Components. Some electronic components such as capacitors or integrated circuits have been found to follow an exponential distribution. Early efforts at collecting reliability data assumed a constant failure rate and therefore many reliability handbooks only provide a failure rate estimates for components.

Random Shocks. It is common for the exponential distribution to model the occurrence of random shocks. An example is the failure of a vehicle tire due to puncture from a nail (random shock). The probability of failure in the next mile is independent of how many miles the tire has travelled (memoryless). The probability of failure when the tire is new is the same as when the tire is old (constant failure rate).

In general component life distributions do not have a constant failure rate, for example due to wear or early failures. Therefore, the exponential distribution is often inappropriate to model most life distributions, particularly mechanical components.

Resources	Online: http://www.weibull.com/LifeDataWeb/the_exponential_distribution.htm http://mathworld.wolfram.com/ExponentialDistribution.html http://en.wikipedia.org/wiki/Exponential_distribution http://socr.ucla.edu/htmls/SOCR_Distributions.html (web calc) Books: Balakrishnan, N. & Basu, A.P., 1996. *Exponential Distribution: Theory, Methods and Applications* 1st ed., CRC. Nelson, W.B., 1982. *Applied Life Data Analysis*, Wiley-Interscience.

Relationship to Other Distributions

2-Parameter Exponential Distribution $Exp(t; \mu, \beta)$	Special Case: $$Exp(t; \lambda) = Exp(t; \mu = 0, \beta = \frac{1}{\lambda})$$
Gamma Distribution $Gamma(t; k, \lambda)$	Let $T_1 \dots T_k \sim Exp(\lambda)$ and $T_t = T_1 + T_2 + \dots + T_k$ Then $$T_t \sim Gamma(k, \lambda)$$ The gamma distribution is the probability density function of the sum of k exponentially distributed time random variables sharing the

same constant rate of occurrence, λ. This is a Homogeneous Poisson Process.

Special Case:
$$Exp(t; \lambda) = Gamma(t; k = 1, \lambda)$$

Let
$$T_1, T_2 \dots \sim Exp(t; \lambda)$$

Given
$$time = T_1 + T_2 + \cdots + T_K + T_{K+1} \cdots$$

Then

Poisson Distribution
$$K \sim Pois(\text{k}; \mu = \lambda t)$$

$Pois(k; \mu)$

The Poisson distribution is the probability of observing exactly k occurrences within a time interval [0, t] where the inter-arrival times of each occurrence is exponentially distributed. This is a Homogeneous Poisson Process.

Special Cases:
$$Pois(\text{k} = 1; \mu = \lambda t) = Exp(t; \lambda)$$

Let
$$X \sim Exp(\lambda) \qquad and \qquad Y = X^{1/\beta}$$

Weibull Distribution
Then
$$Y \sim Weibull(\alpha = \lambda^{\frac{-1}{\beta}}, \beta)$$

$Weibull(t; \alpha, \beta)$
Special Case:
$$Exp(t; \lambda) = Weibull\left(t; \alpha = \frac{1}{\lambda}, \beta = 1\right)$$

Let
$$X \sim Exp(\lambda) \qquad and \qquad Y = [X], \qquad Y \text{ is the integer of } X$$

Geometric Distribution
Then
$$Y \sim Geometric(\alpha, \beta)$$

$Geometric(k; p)$

The geometric distribution is the discrete equivalent of the continuous exponential distribution. The geometric distribution is also memoryless.

Rayleigh Distribution
Let
$$X \sim Exp(\lambda) \qquad and \qquad Y = \sqrt{X}$$
Then

$Rayleigh(t; \alpha)$
$$Y \sim Rayleigh(\alpha = \frac{1}{\sqrt{\lambda}})$$

Chi-square
$\chi^2(x; v)$
Special Case:
$$\chi^2(x; v = 2) = Exp\left(x; \lambda = \frac{1}{2}\right)$$

Pareto Distribution
Let
$$Y \sim Pareto(\theta, \alpha) \qquad and \qquad X = \ln(Y/\theta)$$
$Pareto(t; \theta, \alpha)$
Then
$$X \sim Exp(\lambda = \alpha)$$

Logistic
Distribution
$Logistic(\mu, s)$

Let

$$X \sim Exp(\lambda = 1) \qquad and \qquad Y = \ln\left\{\frac{e^{-X}}{1 + e^{-X}}\right\}$$

Then

$$Y \sim Logistic(0,1)$$

(Hastings et al. 2000, p.127):

2.2. Lognormal Continuous Distribution

Probability Density Function - f(t)

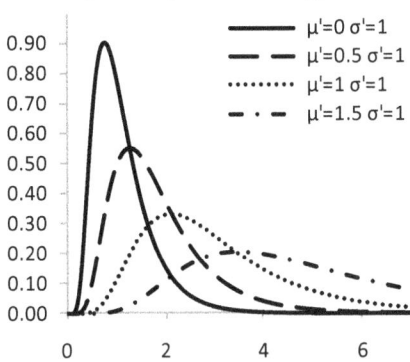

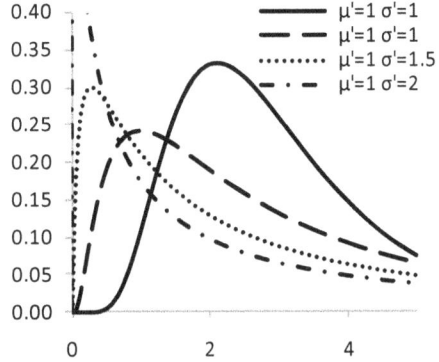

Cumulative Density Function - F(t)

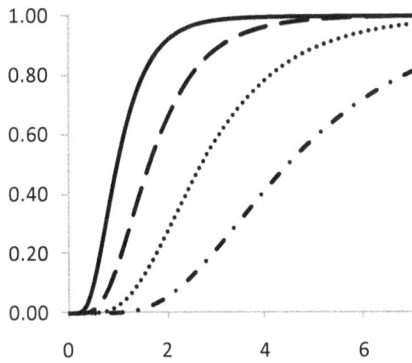

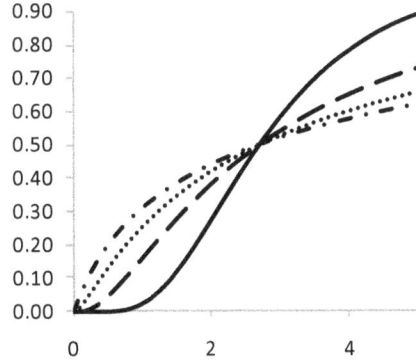

Hazard Rate - h(t)

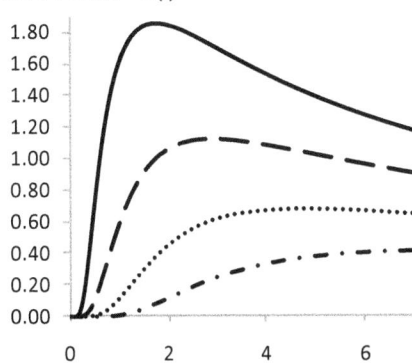

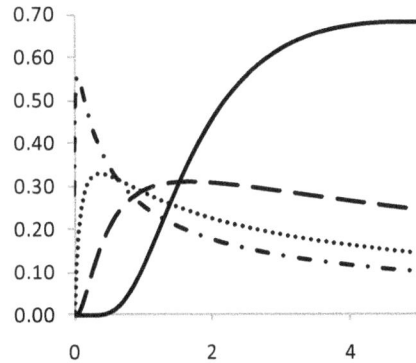

Parameters & Description

Parameters		
	μ_N	$-\infty < \mu_N < \infty$	*Scale parameter:* The mean of the normally distributed $\ln(x)$. This parameter only determines the scale and not the location as in a normal distribution. $$\mu_N = \ln\left(\frac{\mu^2}{\sqrt{\sigma^2 + \mu^2}}\right)$$

Scale parameter: The mean of the normally distributed $\ln(x)$. This parameter only determines the scale and not the location as in a normal distribution.

$$\mu_N = \ln\left(\frac{\mu^2}{\sqrt{\sigma^2 + \mu^2}}\right)$$

$\sigma_N^2 \qquad \sigma_N^2 > 0$

Shape parameter: The standard deviation of the normally distributed ln(x). This parameter only determines the shape and not the scale as in a normal distribution.

$$\sigma_N^2 = \ln\left(\frac{\sigma^2 + \mu^2}{\mu^2}\right)$$

Limits $\qquad t > 0$

Distribution	Formulas

PDF

$$f(t) = \frac{1}{\sigma_N t \sqrt{2\pi}} \exp\left[-\frac{1}{2}\left(\frac{\ln(t) - \mu_N}{\sigma_N}\right)^2\right]$$

$$= \frac{1}{\sigma_N . t} \phi\left[\frac{ln(t) - \mu_N}{\sigma_N}\right]$$

where ϕ is the standard normal pdf.

CDF

$$F(t) = \frac{1}{\sigma_N \sqrt{2\pi}} \int_0^t \frac{1}{t^*} \exp\left[-\frac{1}{2}\left(\frac{\ln(t^*) - \mu_N}{\sigma_N}\right)^2\right] dt^*$$

where, t^* is the time variable.

$$= \frac{1}{2} + \frac{1}{2} \text{erf}\left(\frac{\ln(t) - \mu_N}{\sigma_N \sqrt{2}}\right)$$

$$= \Phi\left(\frac{\ln(t) - \mu_N}{\sigma_N}\right)$$

where Φ is the standard normal cdf.

Reliability

$$R(t) = 1 - \Phi\left(\frac{\ln(t) - \mu_N}{\sigma_N}\right)$$

Conditional Survivor Function $P(T > x + t | T > t)$

$$m(x) = R(x|t) = \frac{R(t + x)}{R(t)} = \frac{1 - \Phi\left(\frac{\ln(x + t) - \mu_N}{\sigma_N}\right)}{1 - \Phi\left(\frac{\ln(t) - \mu_N}{\sigma_N}\right)}$$

Where

t is the given time we know the component has survived to.
x is a random variable defined as the time after t. Note: $x = 0$ at t.

Mean Residual Life

$$u(t) = \frac{\int_t^\infty R(x)dx}{R(t)}$$

$$\lim_{t\to\infty} u(t) \approx \frac{\sigma_N^2 t}{\ln(t) - \mu_N}[1 + o(1)]$$

Where $o(1)$ is Landau's notation. (Kleiber & Kotz 2003, p.114)

Hazard Rate

$$h(t) = \frac{\phi\left[\frac{\ln(t) - \mu_N}{\sigma_N}\right]}{t.\sigma_N\left(1 - \Phi\left[\frac{\ln(t) - \mu_N}{\sigma_N}\right]\right)}$$

Cumulative Hazard Rate

$$H(t) = -\ln[R(t)]$$

Properties and Moments

Median	$e^{(\mu_N)}$
Mode	$e^{(\mu_N - \sigma_N^2)}$
Mean - 1st Raw Moment	$e^{\left(\mu_N + \frac{\sigma_N^2}{2}\right)}$
Variance - 2nd Central Moment	$\left(e^{\sigma_N^2} - 1\right).e^{2\mu_N + \sigma_N^2}$
Skewness - 3rd Central Moment	$\left(e^{\sigma^2} + 2\right).\sqrt{e^{\sigma^2} - 1}$
Excess kurtosis - 4th Central Moment	$e^{4\sigma_N^2} + 2e^{3\sigma_N^2} + 3e^{2\sigma_N^2} - 3$
Characteristic Function	Deriving a unique characteristic equation is not trivial and complex series solutions have been proposed. (Leipnik 1991)
100α% Percentile Function	$t_\alpha = e^{(\mu_N + z_\alpha.\sigma_N)}$

where z_α is the 100pth of the standard normal distribution

$$t_\alpha = e^{(\mu_N + \sigma_N \Phi^{-1}(\alpha))}$$

Parameter Estimation

Plotting Method

Least Square $y = mx + c$	X-Axis	Y-Axis	
Mean	$\ln(t_i)$	$invNorm[F(t_i)]$	$\widehat{\mu_N} = -\frac{c}{m}$ $\widehat{\sigma_N} = \frac{1}{m}$

Maximum Likelihood Function

Likelihood Functions

$$\underbrace{\prod_{i=1}^{n_F} \frac{1}{\sigma_N \cdot t_i^F} \phi(z_i^F)}_{\text{failures}} \cdot \underbrace{\prod_{i=1}^{n_S} [1 - \Phi(z_i^S)]}_{\text{survivors}} \cdot \underbrace{\prod_{i=1}^{n_I} [\Phi(z_i^{RI}) - \Phi(z_i^{LI})]}_{\text{interval failures}}$$

where

$$z_i^x = \left(\frac{\ln(t_i^x) - \mu_N}{\sigma_N} \right)$$

Log-Likelihood Function

$$\Lambda(\mu_N, \sigma_N | E) = \underbrace{\sum_{i=1}^{n_F} \ln\left[\frac{1}{\sigma_N \cdot t_i^F} \phi(z_i^F) \right]}_{\text{failures}} + \underbrace{\sum_{i=1}^{n_S} \ln[1 - \Phi(z_i^S)]}_{\text{survivors}}$$

$$+ \underbrace{\sum_{i=1}^{n_I} \ln[\Phi(z_i^{RI}) - \Phi(z_i^{LI})]}_{\text{interval failures}}$$

where

$$z_i^x = \left(\frac{\ln(t_i^x) - \mu_N}{\sigma_N} \right)$$

$\dfrac{\partial \Lambda}{\partial \mu_N} = 0$

solve for μ_N to get MLE $\widehat{\mu_N}$:

$$\frac{\partial \Lambda}{\partial \mu_N} = \underbrace{\frac{-\mu_N \cdot N^F}{\sigma_N} + \frac{1}{\sigma_N} \sum_{i=1}^{n_F} \ln(t_i^F)}_{\text{failures}} + \underbrace{\frac{1}{\sigma_N} \sum_{i=1}^{n_S} \frac{\phi(z_i^S)}{1 - \Phi(z_i^S)}}_{\text{survivors}}$$

$$- \underbrace{\sum_{i=1}^{n_I} \frac{1}{\sigma_N} \left(\frac{\phi(z_i^{RI}) - \phi(z_i^{LI})}{\Phi(z_i^{RI}) - \Phi(z_i^{LI})} \right)}_{\text{interval failures}} = 0$$

where

$$z_i^x = \left(\frac{\ln(t_i^x) - \mu_N}{\sigma_N} \right)$$

$\dfrac{\partial \Lambda}{\partial \sigma_N} = 0$

solve for σ_N to get $\widehat{\sigma_N}$:

$$\frac{\partial \Lambda}{\partial \sigma_N} = \underbrace{\frac{-n_F}{\sigma_N} + \frac{1}{\sigma_N^3} \sum_{i=1}^{n_F} (\ln(t_i^F) - \mu_N)^2}_{\text{failures}} + \underbrace{\frac{1}{\sigma_N} \sum_{i=1}^{n_S} \frac{z_i^S \cdot \phi(z_i^S)}{1 - \Phi(z_i^S)}}_{\text{survivors}}$$

$$- \underbrace{\sum_{i=1}^{n_I} \frac{1}{\sigma_N} \left(\frac{z_i^{RI} \cdot \phi(z_i^{RI}) - z_i^{LI} \phi(z_i^{LI})}{\Phi(z_i^{RI}) - \Phi(z_i^{LI})} \right)}_{\text{interval failures}} = 0$$

where

$$z_i^x = \left(\frac{\ln(t_i^x) - \mu_N}{\sigma_N} \right)$$

MLE Point Estimates

When there is only complete failure data the point estimates can be given as:

$$\widehat{\mu_N} = \frac{\sum \ln(t_i^F)}{n_F} \qquad \widehat{\sigma_N^2} = \frac{\sum (\ln(t_i^F) - \widehat{\mu_t})^2}{n_F}$$

Note: In almost all cases the MLE methods for a normal distribution can be used by taking the $ln(X)$. However Normal distribution estimation methods cannot be used with interval data. (Johnson et al. 1994, p.220)

In most cases the unbiased estimators are used:

$$\widehat{\mu_N} = \frac{\sum \ln(t_i^F)}{n_F} \qquad \widehat{\sigma_N^2} = \frac{\sum \left(\ln(t_i^F) - \widehat{\mu_t}\right)^2}{n_F - 1}$$

Fisher Information

$$I(\mu_N, \sigma_N^2) = \begin{bmatrix} \dfrac{1}{\sigma_N^2} & 0 \\ 0 & -\dfrac{1}{2\sigma^4} \end{bmatrix}$$

(Kleiber & Kotz 2003, p.119).

$100\gamma\%$ Confidence Intervals

(for complete data)

	1-Sided Lower	2-Sided Lower	2-Sided Upper
μ_N	$\widehat{\mu_N} - \dfrac{\widehat{\sigma_N}}{\sqrt{n_F}} t_\gamma(n_F - 1)$	$\widehat{\mu_N} - \dfrac{\widehat{\sigma_N}}{\sqrt{n_F}} t_{\left(\frac{1-\gamma}{2}\right)}(n_F - 1)$	$\widehat{\mu_N} + \dfrac{\widehat{\sigma_N}}{\sqrt{n_F}} t_{\left(\frac{1-\gamma}{2}\right)}(n_F - 1)$
σ_N^2	$\widehat{\sigma_N^2} \dfrac{(n_F - 1)}{\chi_\gamma^2(n_F - 1)}$	$\widehat{\sigma_N^2} \dfrac{(n_F - 1)}{\chi_{\left(\frac{1+\gamma}{2}\right)}^2(n_F - 1)}$	$\widehat{\sigma_N^2} \dfrac{(n_F - 1)}{\chi_{\left(\frac{1-\gamma}{2}\right)}^2(n_F - 1)}$

Where $t_\gamma(n_F - 1)$ is the $100\gamma^{th}$ percentile of the t-distribution with $n_F - 1$ degrees of freedom and $\chi_\gamma^2(n_F - 1)$ is the $100\gamma^{th}$ percentile of the χ^2-distribution with $n_F - 1$ degrees of freedom. (Nelson 1982, pp.218-219)

	1 Sided - Lower	2 Sided
μ	$\exp\left\{\widehat{\mu_N} + \dfrac{\widehat{\sigma_N^2}}{2} - Z_{1-\alpha}\sqrt{\dfrac{\widehat{\sigma_N^2}}{n_F} + \dfrac{\widehat{\sigma_N^4}}{2(n_F - 1)}}\right\}$	$\exp\left\{\widehat{\mu_N} + \dfrac{\widehat{\sigma_N^2}}{2} \pm Z_{1-\alpha/2}\sqrt{\dfrac{\widehat{\sigma_N^2}}{n_F} + \dfrac{\widehat{\sigma_N^4}}{2(n_F - 1)}}\right\}$

These formulas are the Cox approximation for the confidence intervals of the lognormal distribution mean where $Z_p = \Phi^{-1}(p)$, the inverse of the standard normal cdf. (Zhou & Gao 1997)

Zhou & Gao recommend using the parametric bootstrap method for small sample sizes. (Angus 1994)

Bayesian

Non-informative Priors when σ_N^2 is known, $\pi_0(\mu_N)$
(Yang and Berger 1998, p.22)

Type	Prior	Posterior
Uniform Proper Prior with limits $\mu_N \in [a, b]$	$\dfrac{1}{b - a}$	Truncated Normal Distribution For $a \le \mu_N \le b$ $$c.Norm\left(\mu_N; \frac{\sum_{i=1}^{n_F} \ln t_i^F}{n_F}, \frac{\sigma_N^2}{n_F}\right)$$ Otherwise $\pi(\mu_N) = 0$

| All | 1 | $Norm\left(\mu_N; \dfrac{\sum_{i=1}^{n_F} \ln t_i^F}{n_F}, \dfrac{\sigma_N^2}{n_F}\right)$ |

when $\mu_N \in (\infty, \infty)$

Non-informative Priors when μ_N is known, $\pi_o(\sigma_N^2)$
(Yang and Berger 1998, p.23)

Type	Prior	Posterior
Uniform Proper Prior with limits $\sigma_N^2 \in [a, b]$	$\dfrac{1}{b-a}$	Truncated Inverse Gamma Distribution For $a \le \sigma_N^2 \le b$ $c.\,IG\left(\sigma_N^2; \dfrac{(n_F-2)}{2}, \dfrac{S_N^2}{2}\right)$ Otherwise $\pi(\sigma_N^2) = 0$
Uniform Improper Prior with limits $\sigma_N^2 \in (0, \infty)$	1	$IG\left(\sigma_N^2; \dfrac{(n_F-2)}{2}, \dfrac{S_N^2}{2}\right)$ See section 1.7.1
Jeffery's, Reference, MDIP Prior	$\dfrac{1}{\sigma_N^2}$	$IG\left(\sigma_N^2; \dfrac{n_F}{2}, \dfrac{S_N^2}{2}\right)$ with limits $\sigma_N^2 \in (0, \infty)$ See section 1.7.1

Non-informative Priors when μ_N and σ_N^2 are unknown, $\pi_o(\mu_N, \sigma_N^2)$
(Yang and Berger 1998, p.23)

Type	Prior	Posterior
Improper Uniform with limits: $\mu_N \in (\infty, \infty)$ $\sigma_N^2 \in (0, \infty)$	1	$\pi(\mu_N\|E) \sim T\left(\mu_N; n_F-3, \overline{t_N}, \dfrac{S_N^2}{n_F(n_F-3)}\right)$ See section 1.7.2 $\pi(\sigma_N^2\|E) \sim IG\left(\sigma_N^2; \dfrac{(n_F-3)}{2}, \dfrac{S_N^2}{2}\right)$ See section 1.7.1
Jeffery's Prior	$\dfrac{1}{\sigma_N^4}$	$\pi(\mu_N\|E) \sim T\left(\mu_N; N^F+1, \overline{t_N}, \dfrac{S^2}{n_F(n_F+1)}\right)$ when $\mu_N \in (\infty, \infty)$ See section 1.7.2 $\pi(\sigma_N^2\|E) \sim IG\left(\sigma_N^2; \dfrac{(n_F+1)}{2}, \dfrac{S_N^2}{2}\right)$ when $\sigma_N^2 \in (0, \infty)$ See section 1.7.1
Reference Prior ordering $\{\phi, \sigma\}$	$\pi_o(\phi, \sigma_N^2)$ $\propto \dfrac{1}{\sigma_N\sqrt{2+\phi^2}}$ where $\phi = \mu_N/\sigma_N$	No closed form

Reference where μ and σ^2 are separate groups.

MDIP Prior

$$\frac{1}{\sigma_N}$$

$$\pi(\mu_N|E) \sim T\left(\mu_N; N^F - 1, \overline{t_N}, \frac{S_N^2}{n_F(n_F - 1)}\right)$$

when $\mu_N \in (\infty, \infty)$
See section 1.7.2

$$\pi(\sigma_N^2|E) \sim IG\left(\sigma_N^2; \frac{(n_F - 1)}{2}, \frac{S_N^2}{2}\right)$$

when $\sigma_N^2 \in (0, \infty)$
See section 1.7.1

where

$$S_N^2 = \sum_{i=1}^{n_F} (\ln t_i - \overline{t_N})^2 \quad \text{and} \quad \overline{t_N} = \frac{1}{n_F} \sum_{i=1}^{n_F} \ln t_i$$

Conjugate Priors

UOI	Likelihood Model	Evidence	Dist. of UOI	Prior Para	Posterior Parameters
σ_N^2 from $LogN(t; \mu_N, \sigma_N^2)$	Lognormal with known μ_N	n_F failures at times t_i	Gamma	k_0, λ_0	$k = k_o + n_F/2$ $\lambda = \lambda_o + \frac{1}{2}\sum_{i=1}^{n_F}(\ln t_i - \mu_N)^2$
μ_N from $LogN(t; \mu_N, \sigma_N^2)$	Lognormal with known σ_N^2	n_F failures at times t_i	Normal	μ_o, σ_o^2	$\mu = \dfrac{\frac{\mu_0}{\sigma_0^2} + \frac{\sum_{i=1}^{n_F}\ln(t_i)}{\sigma_N^2}}{\frac{1}{\sigma_0^2} + \frac{n_F}{\sigma_N^2}}$ $\sigma^2 = \dfrac{1}{\frac{1}{\sigma_0^2} + \frac{n_F}{\sigma_N^2}}$

Description, Limitations and Uses

Example

5 components are put on a test with the following failure times:
98, 116, 2485, 2526, , 2920 hours

Taking the natural log of these failure times allows us to use a normal distribution to approximate the parameters. $\ln(t_i)$:
4.590, 4.752, 7.979, 7.818, 7.834 ln(hours)

MLE Estimates are:

$$\widehat{\mu_N} = \frac{\sum \ln(t_i^F)}{n_F} = \frac{32.974}{5} = 6.595$$

$$\widehat{\sigma_N^2} = \frac{\sum(\ln(t_i^F) - \widehat{\mu_t})^2}{n_F - 1} = 3.091$$

90% confidence interval for μ_N:

$$\left[\widehat{\mu_N} - \frac{\widehat{\sigma_N}}{\sqrt{4}} t_{\{0.95\}}(4), \quad \widehat{\mu_N} + \frac{\widehat{\sigma_N}}{\sqrt{4}} t_{\{0.95\}}(4)\right]$$

$$[4.721, \ 8.469]$$

90% confidence interval for σ_N^2:

$$\left[\hat{\sigma_N^2} \frac{4}{\chi_{\{0.95\}}^2(4)}, \quad \hat{\sigma_N^2} \frac{4}{\chi_{\{0.05\}}^2(4)} \right]$$
$$[1.303, \ 17.396]$$

A Bayesian point estimate using the Jeffery non-informative improper prior $1/\sigma_N^4$ with posterior for $\mu_N \sim T(6, \ 6.595, \ 0.412)$ and $\sigma_N^2 \sim IG(3, 6.182)$ has point estimates:

$$\hat{\mu_N} = E[T(6,6.595,0.412)] = \mu = 6.595$$

$$\hat{\sigma_N^2} = E[IG(3,6.182)] = \frac{6.182}{2} = 3.091$$

With 90% confidence intervals:

μ_N
$$[F_T^{-1}(0.05) = 5.348, \quad F_T^{-1}(0.95) = 7.842]$$

σ_N^2
$$[1/F_G^{-1}(0.95) = 0.982, \quad 1/F_G^{-1}(0.05) = 7.560]$$

Characteristics

μ_N **Characteristics.** μ_N determines the scale and not the location as in a normal distribution. The distribution if fixed at f(0)=0 and an increase in the scale parameter stretches the distribution across the x-axis. This has the effect of increasing the mode, mean and median of the distribution.

σ_N **Characteristics.** σ_N determines the shape and not the scale as in a normal distribution. For values of $\sigma_N > 1$ the distribution rises very sharply at the beginning and decreases with a shape similar to an Exponential or Weibull with $0 < \beta < 1$. As $\sigma_N \to 0$ the mode, mean and median converge to e^{μ_N}. The distribution becomes narrower and approaches a Dirac delta function at $t = e^{\mu_N}$.

Hazard Rate. (Kleiber & Kotz 2003, p.115)The hazard rate is unimodal with $h(0) = 0$ and all dirivitives of $h'(t) = 0$ and a slow decrease to zero as $t \to 0$. The mode of the hazard rate:
$$t_m = \exp(\mu + z_m \sigma)$$
where z_m is given by:
$$(z_m + \sigma_N) = \frac{\phi(z_m)}{1 - \Phi(z_m)}$$
therefore $-\sigma_N < z_m < -\sigma_N + \sigma^{-1}$ and therefore:

$$e^{\mu_N - \sigma_N^2} < t_m < e^{\mu_N - \sigma_N^2 + 1}$$
As $\sigma_N \to \infty, t_m \to e^{\mu_N - \sigma_N^2}$ and so for large σ_N:
$$\max h(t) \approx \frac{\exp\left(\mu_N - \frac{1}{2}\sigma_N^2\right)}{\sigma_N \sqrt{2\pi}}$$

As $\sigma_N \to 0, t_m \to e^{\mu_N - \sigma_N^2 + 1}$ and so for large σ_N:

$$\max h(t) \approx \frac{1}{\sigma_N^2 e^{\mu_N - \sigma_N^2 + 1}}$$

Mean / Median / Mode:

$$mode(X) < median(X) < E[X]$$

Scale/Product Property:

Let:

$$a_j X_j \sim LogN(\mu_{Nj}, \sigma_{Nj}^2)$$

If X_j and X_{j+1} are independent:

$$\prod a_j X_j \sim LogN\left(\sum\{\mu_{Nj} + \ln(a_j)\}, \sum \sigma_{Nj}^2\right)$$

Lognormal versus Weibull. In analyzing life data to these distributions, it is often the case that both may be a good fit, especially in the middle of the distribution. The Weibull distribution has an earlier lower tail and produces a more pessimistic estimate of the component life. (Nelson 1990, p.65)

Applications

General Life Distributions. The lognormal distribution has been found to accurately model many life distributions and is a popular choice for life distributions. The increasing hazard rate in early life models the weaker subpopulation (burn in) and the remaining decreasing hazard rate describes the main population. In particular this has been applied to some electronic devices and fatigue-fracture data. (Meeker & Escobar 1998, p.262)

Failure Modes from Multiplicative Errors. The lognormal distribution is very suitable for failure processes that are a result of multiplicative errors. Specific applications include failure of components due to fatigue cracks. (Provan 1987)

Repair Times. The lognormal distribution has commonly been used to model repair times. It is natural for a repair time probability to increase quickly to a mode value. For example, very few repairs have an immediate or quick fix. However, once the time of repair passes the mean it is likely that there are serious problems, and the repair will take a substantial amount of time.

Parameter Variability. The lognormal distribution can be used to model parameter variability. This was done when estimating the uncertainty in the parameter λ in a Nuclear Reactor Safety Study (NUREG-75/014).

Theory of Breakage. The distribution models particle sizes observed in breakage processes (Crow & Shimizu 1988)

Resources   Online:
http://www.weibull.com/LifeDataWeb/the_lognormal_distribution.ht
m
http://mathworld.wolfram.com/LogNormalDistribution.html
http://en.wikipedia.org/wiki/Log-normal_distribution
http://socr.ucla.edu/htmls/SOCR_Distributions.html (web calc)

Books:
Crow, E.L. & Shimizu, K., 1988. *Lognormal distributions*, CRC
Press.

Aitchison, J.J. & Brown, J., 1957. The *Lognormal Distribution*, New
York: Cambridge University Press.

Nelson, W.B., 1982. *Applied Life Data Analysis*, Wiley-Interscience.

Relationship to Other Distributions

Normal
Distribution

$Norm(t; \mu, \sigma^2)$

Let;
$$X \sim LogN(\mu_N, \sigma_N^2)$$
$$Y = \ln(X)$$
Then:
$$Y \sim Norm(\mu, \sigma^2)$$
Where:
$$\mu_N = \ln\left(\frac{\mu^2}{\sqrt{\sigma^2 + \mu^2}}\right), \quad \sigma_N = \ln\left(\frac{\sigma^2 + \mu^2}{\mu^2}\right)$$

2.3. **Weibull Continuous Distribution**

Probability Density Function - f(t)

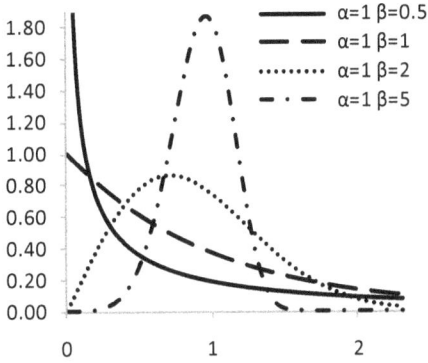

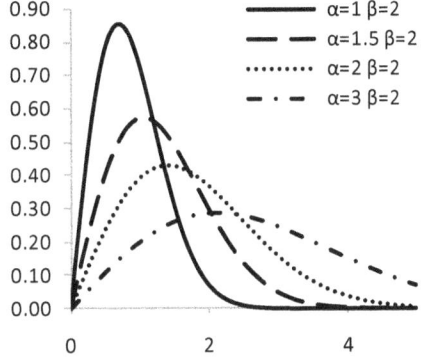

Cumulative Density Function - F(t)

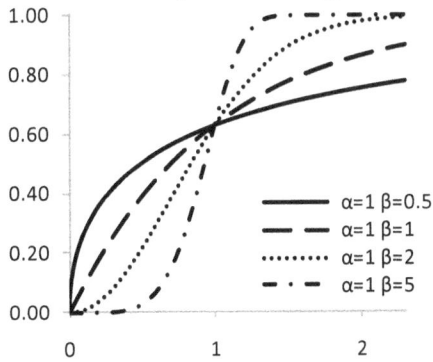

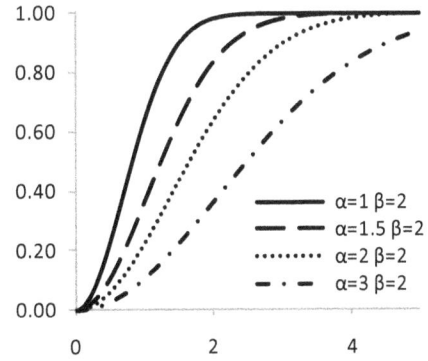

Hazard Rate - h(t)

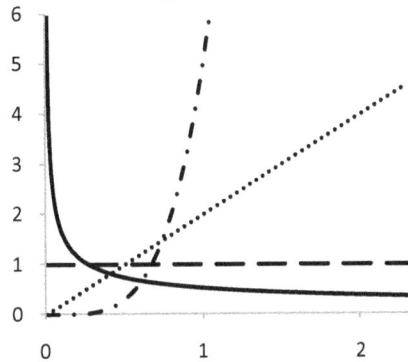

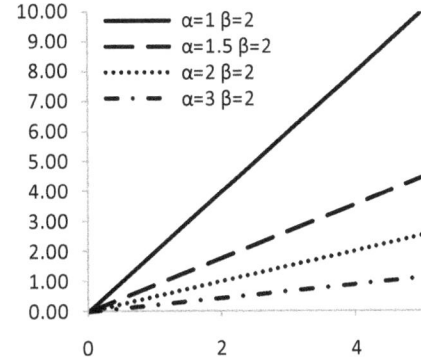

Parameters & Description

Parameters			
	α	$\alpha > 0$	*Scale Parameter:* The value of α equals the 63.2th percentile and has a unit equal to t. Note that this is not equal to the mean.
	β	$\beta > 0$	*Shape Parameter:* Also known as the slope (referring to a linear CDF plot) β determines the shape of the distribution.
Limits		$t \geq 0$	

Distribution	Formulas
PDF	$$f(t) = \frac{\beta t^{\beta-1}}{\alpha^\beta} e^{-\left(\frac{t}{\alpha}\right)^\beta}$$
CDF	$$F(t) = 1 - e^{-\left(\frac{t}{\alpha}\right)^\beta}$$
Reliability	$$R(t) = e^{-\left(\frac{t}{\alpha}\right)^\beta}$$

Conditional Survivor Function
$P(T > x + t | T > t)$

$$m(x) = R(x|t) = \frac{R(t+x)}{R(t)} = e^{\left(\frac{t^\beta - (t+x)^\beta}{\alpha^\beta}\right)}$$

Where
t is the given time we know the component has survived to.
x is a random variable defined as the time after t. Note: $x = 0$ at t.

(Kleiber & Kotz 2003, p.176)

Mean Residual Life

$$u(t) = e^{\left(\frac{t}{\alpha}\right)^\beta} \int_t^\infty e^{-\left(\frac{x}{\alpha}\right)^\beta} dx$$

which has the asymptotic property of:
$$\lim_{t \to \infty} u(t) = t^{1-\beta}$$

Hazard Rate	$$h(t) = \frac{\beta}{\alpha} \cdot \left(\frac{t}{\alpha}\right)^{\beta-1}$$
Cumulative Hazard Rate	$$H(t) = \left(\frac{t}{\alpha}\right)^\beta$$

Properties and Moments

Median	$$\alpha\left(ln(2)\right)^{\frac{1}{\beta}}$$
Mode	$$\alpha\left(\frac{\beta-1}{\beta}\right)^{\frac{1}{\beta}} \quad \text{if } \beta \geq 1$$ otherwise no mode exists

Mean - 1st Raw Moment	$\alpha \Gamma \left(1 + \dfrac{1}{\beta}\right)$
Variance - 2nd Central Moment	$\alpha^2 \left[\Gamma\left(1 + \dfrac{2}{\beta}\right) - \Gamma^2\left(1 + \dfrac{1}{\beta}\right)\right]$
Skewness - 3rd Central Moment	$\dfrac{\Gamma\left(1 + \dfrac{3}{\beta}\right)\alpha^3 - 3\mu\sigma^2 - \mu^3}{\sigma^3}$
Excess kurtosis - 4th Central Moment	$\dfrac{-6\Gamma_1^4 + 12\Gamma_1^2\Gamma_2 - 3\Gamma_2^2 - 4\Gamma_1\Gamma_3 + \Gamma_4}{(\Gamma_2 - \Gamma_1^2)^2}$

where:

$$\Gamma_i = \Gamma\left(1 + \frac{i}{\beta}\right)$$

Characteristic Function	$\displaystyle\sum_{n=0}^{\infty} \dfrac{(it)^n \alpha^n}{n!} \Gamma\left(1 + \dfrac{n}{\beta}\right)$
100p% Percentile Function	$t_p = \alpha[-\ln(1-p)]^{\frac{1}{\beta}}$

Parameter Estimation

Plotting Method

Least Square $y = mx + c$	Mean	X-Axis $\ln(t_i)$	Y-Axis $\ln\left[ln\left(\dfrac{1}{1-F}\right)\right]$	$\hat{\alpha} = e^{-\frac{c}{m}}$ $\hat{\beta} = m$

Maximum Likelihood Function

Likelihood Functions	$L(\alpha,\beta|E) = \underbrace{\prod_{i=1}^{n_F} \dfrac{\beta\left(t_i^F\right)^{\beta-1}}{\alpha^\beta} e^{-\left(\frac{t_i^F}{\alpha}\right)^\beta}}_{\text{failures}} \cdot \underbrace{\prod_{i=1}^{n_S} e^{-\left(\frac{t_i^S}{\alpha}\right)^\beta}}_{\text{survivors}}$ $\underbrace{\prod_{i=1}^{n_I} \left(e^{-\left(\frac{t_i^{LI}}{\alpha}\right)^\beta} - e^{-\left(\frac{t_i^{RI}}{\alpha}\right)^\beta} \right)}_{\text{interval failures}}$
Log-Likelihood Function	$\Lambda(\alpha,\beta|E) = n_F \ln(\beta) - \beta n_F \ln(\alpha) + \underbrace{\sum_{i=1}^{n_F}\left\{(\beta-1)\ln\left(t_i^F\right) - \left(\dfrac{t_i^F}{\alpha}\right)^\beta\right\}}_{\text{failures}}$ $\underbrace{-\sum_{i=1}^{n_S}\left(\dfrac{t_i^S}{\alpha}\right)^\beta}_{\text{survivors}} + \underbrace{\sum_{i=1}^{n_I} \ln\left(e^{-\left(\frac{t_i^{LI}}{\alpha}\right)^\beta} - e^{-\left(\frac{t_i^{RI}}{\alpha}\right)^\beta} \right)}_{\text{interval failures}}$

$\dfrac{\partial \Lambda}{\partial \alpha} = 0$ solve for α to get $\hat{\alpha}$:

$$\frac{\partial \Lambda}{\partial \alpha} = \underbrace{\frac{-\beta n_F}{\alpha} + \frac{\beta}{\alpha^{\beta+1}} \sum_{i=1}^{n_F}\left(t_i^F\right)^\beta}_{\text{failures}} + \underbrace{\frac{\beta}{\alpha^{\beta+1}} \sum_{i=1}^{n_S}\left(t_i^S\right)^\beta}_{\text{survivors}}$$

$$+ \underbrace{\sum_{i=1}^{n_I} \frac{\beta}{\alpha} \left(\frac{\left(\frac{t_i^{LI}}{\alpha}\right)^\beta e^{\left(\frac{t_i^{RI}}{\alpha}\right)^\beta} - \left(\frac{t_i^{RI}}{\alpha}\right)^\beta e^{\left(\frac{t_i^{LI}}{\alpha}\right)^\beta}}{e^{\left(\frac{t_i^{RI}}{\alpha}\right)^\beta} - e^{\left(\frac{t_i^{LI}}{\alpha}\right)^\beta}} \right)}_{\text{interval failures}} = 0$$

$\dfrac{\partial \Lambda}{\partial \beta} = 0$ solve for β to get $\hat{\beta}$:

$$\frac{\partial \Lambda}{\partial \beta} = \underbrace{\frac{n_F}{\beta} + \sum_{i=1}^{n_F}\left\{\ln\left(\frac{t_i^F}{\alpha}\right) - \left(\frac{t_i^F}{\alpha}\right)^\beta \cdot \ln\left(\frac{t_i^F}{\alpha}\right)\right\}}_{\text{failures}} - \underbrace{\sum_{i=1}^{n_S}\left(\frac{t_i^S}{\alpha}\right)^\beta \ln\left(\frac{t_i^S}{\alpha}\right)}_{\text{survivors}}$$

$$+ \underbrace{\sum_{i=1}^{n_I} \left(\frac{\ln\left(\frac{t_i^{RI}}{\alpha}\right) \cdot \left(\frac{t_i^{RI}}{\alpha}\right)^\beta \cdot e^{\left(\frac{t_i^{LI}}{\alpha}\right)^\beta} - \ln\left(\frac{t_i^{LI}}{\alpha}\right) \cdot \left(\frac{t_i^{LI}}{\alpha}\right)^\beta \cdot e^{\left(\frac{t_i^{RI}}{\alpha}\right)^\beta}}{e^{\left(\frac{t_i^{RI}}{\alpha}\right)^\beta} - e^{\left(\frac{t_i^{LI}}{\alpha}\right)^\beta}} \right)}_{\text{interval failures}} = 0$$

MLE Point
Estimates

When there is only complete failure and/or right censored data the point estimates can be solved using (Rinne 2008, p.439):

$$\hat{\alpha} = \left[\frac{\Sigma\left(t_i^F\right)^{\hat{\beta}} + \Sigma\left(t_i^S\right)^{\hat{\beta}}}{n_F} \right]^{\frac{1}{\hat{\beta}}}$$

$$\hat{\beta} = \left[\frac{\Sigma\left(t_i^F\right)^{\hat{\beta}} \ln\left(t_i^F\right) + \Sigma\left(t_i^S\right)^{\hat{\beta}} \ln\left(t_i^S\right)}{\Sigma\left(t_i^F\right)^{\hat{\beta}} + \Sigma\left(t_i^S\right)^{\hat{\beta}}} - \frac{1}{n_F}\sum \ln\left(t_i^F\right) \right]^{-1}$$

Note: Numerical methods are needed to solve $\hat{\beta}$ then substitute to find $\hat{\alpha}$. Numerical methods to find Weibull MLE estimates for complete and censored data for 2 parameter and 3 parameter Weibull distribution are detailed in (Rinne 2008).

Fisher
Information
Matrix

(Rinne   2008,
p.412)

$$I(\alpha,\beta) = \begin{bmatrix} \dfrac{\beta^2}{\alpha^2} & \dfrac{\Gamma'(2)}{-\alpha} \\ \dfrac{\Gamma'(2)}{-\alpha} & \dfrac{1 + \Gamma''(2)}{\beta^2} \end{bmatrix} = \begin{bmatrix} \dfrac{\beta^2}{\alpha^2} & \dfrac{1-\gamma}{\alpha} \\ \dfrac{1-\gamma}{\alpha} & \dfrac{\frac{\pi^2}{6} + (1-\gamma^2)}{\beta^2} \end{bmatrix}$$

$$\cong \begin{bmatrix} \dfrac{\beta^2}{\alpha^2} & \dfrac{0.422784}{-\alpha} \\ \dfrac{0.422784}{-\alpha} & \dfrac{1.823680}{\beta^2} \end{bmatrix}$$

100γ% Confidence Interval (complete data)	The asymptotic variance-covariance matrix of $(\hat{\alpha}, \hat{\beta})$ is: (Rinne 2008, pp.412-417)

$$Cov(\hat{\alpha}, \hat{\beta}) = [J_n(\hat{\alpha}, \hat{\beta})]^{-1} = \frac{1}{n_F}\begin{bmatrix} 1.1087\dfrac{\hat{\alpha}^2}{\hat{\beta}^2} & 0.2570\hat{\alpha} \\ 0.2570\hat{\alpha} & 0.6079\hat{\beta}^2 \end{bmatrix}$$

Bayesian

Bayesian analysis is applied to either one of two re-parameterizations of the Weibull Distribution: (Rinne 2008, p.517)

$$f(t; \lambda, \beta) = \lambda\beta t^{\beta-1}\exp(-\lambda t^\beta) \quad \text{where } \lambda = \alpha^{-\beta}$$

or

$$f(t; \theta, \beta) = \frac{\beta}{\theta}t^{\beta-1}\exp\left(-\frac{t^\beta}{\theta}\right) \quad \text{where } \theta = \frac{1}{\lambda} = \alpha^\beta$$

Non-informative Priors $\pi_0(\lambda)$ (Rinne 2008, p.517)

Type	Prior	Posterior
Uniform Proper Prior with known β and limits $\lambda \in [a, b]$	$\dfrac{1}{b-a}$	Truncated Gamma Distribution For $a \le \lambda \le b$ $c.\,Gamma(\lambda; 1 + n_F, t_{T,\beta})$ Otherwise $\pi(\lambda) = 0$
Jeffrey's Prior when β is known.	$\dfrac{1}{\lambda} \propto Gamma(0,0)$	$Gamma(\lambda; n_F, t_{T,\beta})$ when $\lambda \in [0, \infty)$
Jeffrey's Prior for unknown θ and β.	$\dfrac{1}{\theta\beta}$	No closed form (Rinne 2008, p.527)

where $\quad t_{T,\beta} = \Sigma(t_i^F)^\beta + \Sigma(t_i^S)^\beta = $ adjusted total time in test

Conjugate Priors

No joint continuous prior distribution exists for the Weibull distribution. However, a procedure exist which use a continuous distribution for α and a discrete distribution for β which will not be included here. (Martz & Waller 1982)

UOI	Likelihood Model	Evidence	Dist. of UOI	Prior Para	Posterior Parameters
λ where $\lambda = \alpha^{-\beta}$ from $Wbl(t; \alpha, \beta)$	Weibull with known β	n_F failures at times t_i^F	Gamma	k_0, Λ_0	$k = k_0 + n_F$ $\Lambda = \Lambda_0 + t_{T,\beta}$ (Rinne 2008, p.520)

θ where $\theta = \alpha^\beta$ from $Wbl(t; \alpha, \beta)$	Weibull with known β	n_F failures at times t_i^F	Inverted Gamma	α_0, β_0	$\alpha = \alpha_o + n_F$ $\beta = \beta_0 + t_{T,\beta}$ (Rinne 2008, p.524)

Description, Limitations and Uses

Example

5 components are put on a test with the following failure times: 535, 613, 976, 1031, 1875 hours.

$\hat\beta$ is found by numerically solving:

$$\hat\beta = \left[\frac{\Sigma(t_i^F)^{\hat\beta} \ln(t_i^F)}{\Sigma(t_i^F)^{\hat\beta}} - 6.8118 \right]^{-1}$$

$$\hat\beta = 2.275$$

$\hat\alpha$ is found by solving:

$$\hat\alpha = \left[\frac{\Sigma(t_i^F)^{\hat\beta}}{n_F} \right]^{\frac{1}{\hat\beta}} = 1140$$

Covariance Matrix is:

$$Cov(\hat\alpha, \hat\beta) = \frac{1}{5}\begin{bmatrix} 1.1087\frac{\hat\alpha^2}{\hat\beta^2} & 0.2570\hat\alpha \\ 0.2570\hat\alpha & 0.6079\hat\beta^2 \end{bmatrix} = \begin{bmatrix} 55679 & 58.596 \\ 58.596 & 0.6293 \end{bmatrix}$$

90% confidence interval for $\hat\alpha$:

$$\left[\hat\alpha . \exp\left\{ \frac{\Phi^{-1}(0.95)\sqrt{55679}}{-\hat\alpha} \right\}, \quad \hat\alpha . \exp\left\{ \frac{\Phi^{-1}(0.95)\sqrt{55679}}{\hat\alpha} \right\} \right]$$

$$[811, \quad 1602]$$

90% confidence interval for β:

$$\left[\hat\beta . \exp\left\{ \frac{\Phi^{-1}(0.95)\sqrt{0.6293}}{-\hat\beta} \right\}, \quad \hat\beta . \exp\left\{ \frac{\Phi^{-1}(0.95)\sqrt{0.6293}}{\hat\beta} \right\} \right]$$

$$[1.282, \quad 4.037]$$

Note that with only 5 samples the assumption that the parameter distribution is approximately normal is probably inaccurate and therefore these confidence intervals need to be used with caution.

Characteristics

The Weibull distribution is also known as a "Type III asymptotic distribution for minimum values".

β Characteristics:

β < 1. The hazard rate decreases with time.

$\beta = 1$. The hazard rate is constant (expoential distribution)
$\beta > 1$. The hazard rate increases with time.
$1 < \beta < 2$. The hazard rate increases less as time increases.
$\beta = 2$. The hazard rate increases with a linear relationship to time.
$\beta > 2$. The hazard rate increases more as time increases.
$\beta < 3.447798$. The distribution is positively skewed. (Tail to right).
$\beta \approx 3.447798$. The distribution is approximately symmetrical.
$\beta > 3.447798$. The distribution is negatively skewed (Tail to left).
$3 < \beta < 4$. The distribution approximates a normal distribution.
$\beta > 10$. The distribution approximates a Smallest Extreme Value Distribution.

Note that for $\beta = 0.999$, $f(0) = \infty$, but for $\beta = 1.001$, $f(0) = 0$. This rapid change creates complications when maximizing likelihood functions. (Weibull.com) As $\beta \rightarrow \infty$, the $mode \rightarrow \alpha$.

α Characteristics. Increasing α stretches the distribution over the time scale. With the $f(0)$ point fixed this also has the effect of increasing the mode, mean and median. The value for α is at the 63% Percentile. $F(\alpha) = 0.632..$

$$X \sim Weibull(\alpha, \beta)$$

Scaling property: (Leemis & McQueston 2008)

$$kX \sim Weibull(\alpha k^\beta, \beta)$$

Minimum property (Rinne 2008, p.107)

$$\min\{X, X_2, \dots, X_n\} \sim Weibull(\alpha n^{-\frac{1}{\beta}}, \beta)$$

When β is fixed.

Variate Generation property

$$F^{-1}(u) = \alpha[-\ln(1-u)]^{\frac{1}{\beta}}, \quad 0 < u < 1$$

Lognormal versus Weibull. In analyzing life data it is often the case that both distributions may be a good fit, especially in the middle of the distribution. The Weibull distribution has an earlier lower tail and produces a more pessimistic estimate of the component life. (Nelson 1990, p.65)

Applications

The Weibull distribution is by far the most popular life distribution used in reliability engineering. This is due to its variety of shapes and generalization or approximation of many other distributions. Analysis assuming a Weibull distribution already includes the exponential life distribution as a special case.

There are many physical interpretations of the Weibull Distribution. Due to its minimum property a physical interpretation is the weakest link, where a system such as a chain will fail when the weakest link fails. It can also be shown that the Weibull Distribution can be derived from a cumulative wear model (Rinne 2008, p.15)

The following is a non-exhaustive list of applications where the Weibull distribution has been used in:
- Acceptance sampling
- Warranty analysis
- Maintenance and renewal
- Strength of material modeling
- Wear modeling
- Electronic failure modeling
- Corrosion modeling

A detailed list with references to practical examples is contained in (Rinne 2008, p.275)

Resources

Online:
http://www.weibull.com/LifeDataWeb/the_weibull_distribution.htm
http://mathworld.wolfram.com/WeibullDistribution.html
http://en.wikipedia.org/wiki/Weibull_distribution
http://socr.ucla.edu/htmls/SOCR_Distributions.html (interactive web calculator)
http://www.qualitydigest.com/jan99/html/weibull.html (how to use conduct Weibull analysis in Excel, *William W. Dorner*)

Books:
Rinne, H., 2008. The Weibull Distribution: A Handbook 1st ed., Chapman & Hall/CRC.

Murthy, D.N.P., Xie, M. & Jiang, R., 2003. Weibull Models 1st ed., Wiley-Interscience.

Nelson, W.B., 1982. *Applied Life Data Analysis*, Wiley-Interscience.

Relationship to Other Distributions

Three Parameter Weibull Distribution

The three-parameter model adds a locator parameter to the two parameter Weibull distribution allowing a shift along the x-axis. This creates a period of guaranteed zero failures to the beginning of the product life and is therefore only used in special cases.

$Weibull(t; \alpha, \beta, \gamma)$

Special Case:
$$Weibull(t; \alpha, \beta) = Weibull(t; \alpha, \beta, \gamma = 0)$$

Exponential
Distribution

$Exp(t; \lambda)$

Let

$$X \sim Weibull(\alpha, \beta) \qquad and \qquad Y = X^\beta$$

Then

$$Y \sim Exp(\lambda = \alpha^{-\beta})$$

Special Case:

$$Exp(t; \lambda) = Weibull\left(t; \alpha = \frac{1}{\lambda}, \beta = 1\right)$$

Rayleigh
Distribution

$Rayleigh(t; \alpha)$

Special Case:

$$Rayleigh(t; \alpha) = Weibull(t; \alpha, \beta = 2)$$

Chi Distribution

$\chi(t|v)$

Special Case:

$$\chi(t|v = 2) = Weibull\left(t|\alpha = \sqrt{2}, \beta = 2\right)$$

3. Bathtub Life Distributions

3.1. 2-Fold Mixed Weibull Distribution

All shapes shown are variations from $p = 0.5$ $\alpha_1 = 2$ $\beta_1 = 0.5$ $\alpha_2 = 10$ $\beta_2 = 20$

Probability Density Function - f(t)

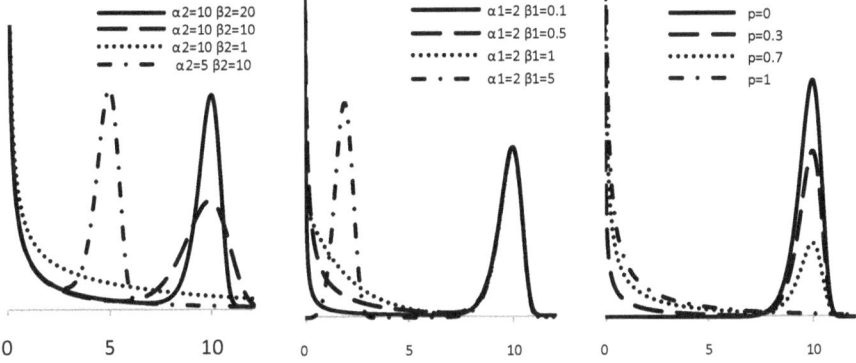

Cumulative Density Function - F(t)

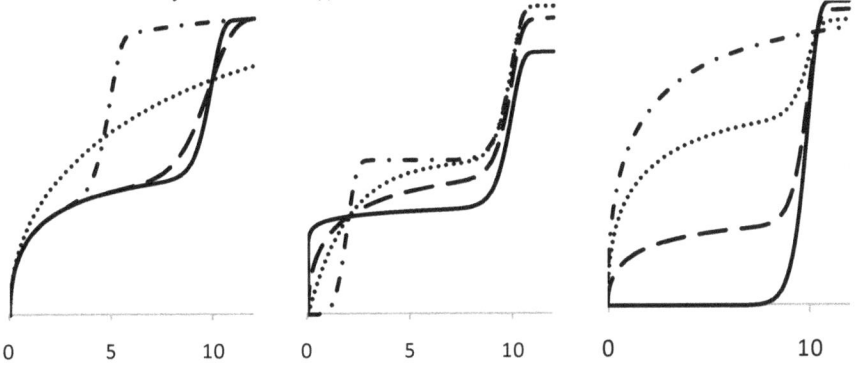

Hazard Rate - h(t)

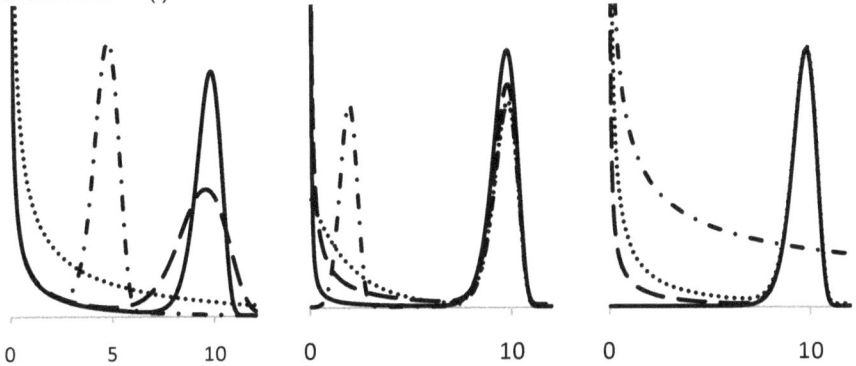

Parameters & Description

Parameters	α_i	$\alpha_i > 0$	*Scale Parameter:* This is the scale for each Weibull Distribution.
	β_i	$\beta_i > 0$	*Shape Parameters:* The shape of each Weibull Distribution
	p	$0 \le p \le 1$	*Mixing Parameter.* This determines the weight each Weibull Distribution has on the overall density function.
Limits		$t \ge 0$	

Distribution / Formulas

PDF

$$f(t) = p f_1(t) + (1-p) f_2(t)$$

$$\text{where } f_i(t) = \frac{\beta_i t^{\beta_i - 1}}{\alpha_i^{\beta_i}} e^{-\left(\frac{t}{\alpha_i}\right)^{\beta_i}} \quad \text{and } i \in \{1,2\}$$

CDF

$$F(t) = p F_1(t) + (1-p) F_2(t)$$

$$\text{where } F_i(t) = 1 - e^{-\left(\frac{t}{\alpha_i}\right)^{\beta_i}} \quad \text{and } i \in \{1,2\}$$

Reliability

$$R(t) = p R_1(t) + (1-p) R_2(t)$$

$$\text{where } R_i(t) = e^{-\left(\frac{t}{\alpha_i}\right)^{\beta_i}} \quad \text{and } i \in \{1,2\}$$

Hazard Rate

$$h(t) = w_1(t) h_1(t) + w_2(t) h_2(t)$$

$$\text{where} \quad w_i(t) = \frac{p_i R_i(t)}{\sum_{i=1}^{n} p_i R_i(t)} \quad \text{and } i \in \{1,2\}$$

Properties and Moments

Median	Solved numerically
Mode	Solved numerically

Mean - 1st Raw Moment

$$p\alpha_1 \Gamma\left(1 + \frac{1}{\beta_1}\right) + (1-p)\alpha_2 \Gamma\left(1 + \frac{1}{\beta_2}\right)$$

Variance - 2nd Central Moment

$$p.Var[T_1] + (1-p)Var[T_2]$$
$$+ p(E[T_1] - E[T])^2$$
$$+ (1-p)(E[T_2] - E[T])^2$$

$$p.\alpha^2\left[\Gamma\left(1+\tfrac{2}{\beta_1}\right)-\Gamma^2\left(1+\tfrac{1}{\beta_1}\right)\right]$$
$$+(1-p)\alpha^2\left[\Gamma\left(1+\tfrac{2}{\beta_2}\right)-\Gamma^2\left(1+\tfrac{1}{\beta_2}\right)\right]$$
$$+p\left[\alpha_1\Gamma\left(1+\tfrac{1}{\beta_1}\right)-E[T]\right]^2$$
$$+(1-p)\left[\alpha_2\Gamma\left(1+\tfrac{1}{\beta_2}\right)-E[T]\right]^2$$

100p% Percentile Function   Solved numerically

Parameter Estimation

Plotting Method (Jiang & Murthy 1995)

Plot Points on a Weibull Probability Plot	X-Axis	Y-Axis
	$x = ln(t)$	$y = ln\left[ln\left(\dfrac{1}{1-F}\right)\right]$

Using the Weibull Probability Plot the parameters can be estimated. Jiang & Murthy, 1995, provide a comprehensive coverage of this procedure and detail error in previous methods. A typical WPP for a 2-fold Mixed Weibull Distribution is:

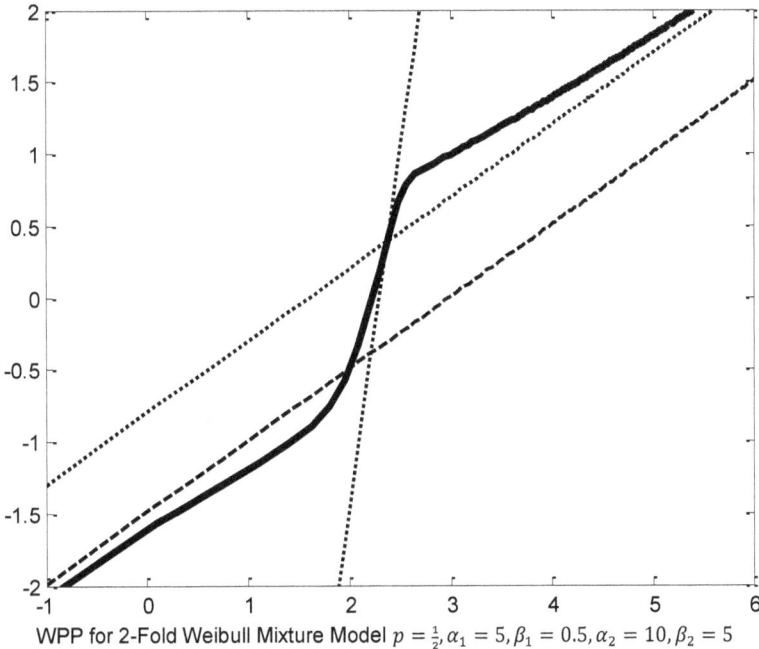

WPP for 2-Fold Weibull Mixture Model $p = \tfrac{1}{2}, \alpha_1 = 5, \beta_1 = 0.5, \alpha_2 = 10, \beta_2 = 5$

Sub Populations:
The dotted lines in the WPP is the lines representing the subpopulations:
$$L_1 = \beta_1[x - \ln(\alpha_1)]$$

$$L_2 = \beta_2[x - \ln(\alpha_2)]$$

Asymptotes (Jiang & Murthy 1995):
As $x \to -\infty$ ($t \to 0$) there exists an asymptote approximated by:
$$y \approx \beta_1[x - \ln(\alpha_1)] + \ln(c)$$
where

$$c = \begin{cases} p & \text{when } \beta_1 \neq \beta_2 \\ p + (1-p) \cdot \left(\dfrac{\alpha_1}{\alpha_2}\right)^{\beta_1} & \text{when } \beta_1 = \beta_2 \end{cases}$$

As $x \to \infty$ ($t \to \infty$) the asymptote straight line can be approximated by:
$$y \approx \beta_1[x - \ln(\alpha_1)]$$

Parameter Estimation
Jiang and Murthy divide the parameter estimation procedure into three cases:

Well Mixed Case $\beta_2 \neq \beta_1$ and $\alpha_1 \approx \alpha_2$
- Estimate the parameters of α_1 and β_1 from the L_1 line (right asymptote).
- Estimate the parameter p from the separation distance between the left and right asymptotes.
- Find the point where the curve crosses L_1 (point I). The slope at point I is:
$$\bar{\beta} = p\beta_1 + (1-p)\beta_2$$
- Determine slope at point I and use to estimate β_2
- Draw a line through the intersection point I with slope β_2 and use the intersection point to estimate α_2.

Well Separated Case $\beta_2 \neq \beta_1$ and $\alpha_1 \gg \alpha_2$ or $\alpha_1 \ll \alpha_2$
- Determine visually if data is scattered along the bottom (or top) to determine if $\alpha_1 \ll \alpha_2$ (or $\alpha_1 \gg \alpha_2$).
- If $\alpha_1 \ll \alpha_2$ ($\alpha_1 \gg \alpha_2$) locate the inflection, y_a, to the left (right) of the point I. This point $y_a \cong \ln[-\ln(1-p)]$ { or $y_a \cong \ln[-\ln(p)]$ }. Using this formula estimate p.
- Estimate α_1 and α_2:
  - If $\alpha_1 \ll \alpha_2$ calculate point $y_1 = \ln\left[\ln\left(1 - p + \frac{p}{exp(1)}\right)\right]$ and $y_2 = \ln\left[\ln\left(\frac{1-p}{exp(1)}\right)\right]$. Find the coordinates where y_1 and y_2 intersect the WPP curve. At these points estimate $\alpha_1 = e^{x_1}$ and $\alpha_2 = e^{x_2}$.
  - If $\alpha_1 \gg \alpha_2$ calculate point $y_1 = \ln\left[-\ln\left(\frac{p}{exp(1)}\right)\right]$ and $y_2 = \ln\left[-\ln\left(p + \frac{1-p}{exp(1)}\right)\right]$. Find the coordinates where y_1 and y_2 intersect the WPP curve. At these points estimate $\alpha_1 = e^{x_1}$ and $\alpha_2 = e^{x_2}$.
- Estimate β_1:
  - If $\alpha_1 \ll \alpha_2$ draw and approximate L_2 ensuring it intersects α_2. Estimate β_2 from the slope of L_2.
  - If $\alpha_1 \gg \alpha_2$ draw and approximate L_1 ensuring it intersects α_1. Estimate β_1 from the slope of L_1.
- Find the point where the curve crosses L_1 (point I). The slope at point I is:
$$\bar{\beta} = p\beta_1 + (1-p)\beta_2$$
- Determine slope at point I and use to estimate β_2

Common Shape Parameter $\beta_2 = \beta_1$

If $\left(\frac{\alpha_2}{\alpha_1}\right)^{\beta_1} \approx 1$ then:
- Estimate the parameters of α_1 and β_1 from the L_1 line (right asymptote).
- Estimate the parameter p from the separation distance between the left and right asymptotes.
- Draw a vertical line through $x = \ln(\alpha_1)$. The intersection with the WPP can yield an estimate of α_2 using:

$$y_1 = \left(\frac{p}{\exp(1)} + \frac{1-p}{\exp\left\{\left(\frac{\alpha_2}{\alpha_1}\right)^{\beta_1}\right\}}\right)$$

If $\left(\frac{\alpha_2}{\alpha_1}\right)^{\beta_1} \ll 1$ then:
- Find inflection point and estimate the y coordinate y_r. Estimate p using:
$$y_T \cong \ln[-\ln(p)]$$
- If $\alpha_1 \ll \alpha_2$ calculate point $y_1 = \ln\left[\ln\left(1 - p + \frac{p}{exp(1)}\right)\right]$ and $y_2 = \ln\left[\ln\left(\frac{1-p}{exp(1)}\right)\right]$. Find the coordinates where y_1 and y_2 intersect the WPP curve. At these points estimate $\alpha_1 = e^{x_1}$ and $\alpha_2 = e^{x_2}$.
- Using the left or right asymptote estimate $\beta_1 = \beta_2$ from the slope.

Maximum Likelihood	MLE and Bayesian techniques can be used using numerical methods however estimates obtained from the graphical methods are useful for initial guesses. A literature review of MLE and Bayesian methods is covered in (Murthy, et al. 2003).
Bayesian	

Description, Limitations and Uses

Characteristics	**Hazard Rate Shape.** The hazard rate can be approximated at its limits by (Jiang & Murthy 1995):

$$Small\ t:\ h(t) \approx ch_1(t) \quad Large\ t: h(t) \approx h_1$$

This result proves that the hazard rate (increasing or decreasing) of h_1 will dominate the limits of the mixed Weibull distribution. Therefore, the hazard rate cannot be a bathtub curve shape. Instead the possible shapes of the hazard rate is:
- Decreasing
- Unimodal
- Decreasing followed by unimodal (rollercoaster)
- Bi-modal

The reason this distribution has been included as a bathtub distribution is because on many occasions the hazard rate of a complex product may follow the "rollercoaster" shape instead which is given as decreasing followed by unimodal shape.

The shape of the hazard rate is only determined by the two shape parameters β_1 and β_2. A complete study on the characterization of the 2-Fold Mixed Weibull Distribution is contained in Jiang and Murthy 1998.

p Values
The mixture ratio, p_i, for each Weibull Distribution may be used to estimate the percentage of each subpopulation. However this is not a reliable measure and it known to be misleading (Berger & Sellke 1987)

N-Fold Distribution (Murthy et al. 2003)
A generalization to the 2-fold mixed Weibull distribution is the n-fold case. This distribution is defined as:

$$f(t) = \sum_{i=1}^{n} p_i f_i(t)$$

$$\text{where } f_i(t) = \frac{\beta_i t^{\beta_i-1}}{\alpha_i^{\beta_i}} e^{-\left(\frac{t}{\alpha_i}\right)^{\beta_i}} \text{ and } \sum_{i=1}^{n} p_i = 1$$

and the hazard rate is given as:

$$h(t) = \sum_{i=1}^{n} w_i(t) h_i(t)$$

$$\text{where } \quad w_i(t) = \frac{p_i R_i(t)}{\sum_{i=1}^{n} p_i R_i(t)}$$

It has been found that in many instances a higher number of folds will not significantly increase the accuracy of the model but does impose a significant overhead in the number of parameters to estimate. The 3-Fold Weibull Mixture Distribution has been studied by Jiang and Murthy 1996.

2-Fold Weibull 3-Parameter Distribution
A common variation to the model presented here is to have the second Weibull distribution modeled with three parameters.

Resources

Books / Journals:
Jiang, R. & Murthy, D., 1995. *Modeling Failure-Data by Mixture of 2 Weibull Distributions: A Graphical Approach.* IEEE Transactions on Reliability, 44, 477-488.

Murthy, D., Xie, M. & Jiang, R., 2003. *Weibull Models* 1st ed., Wiley-Interscience.

Rinne, H., 2008. The *Weibull Distribution: A Handbook* 1st ed., Chapman & Hall/CRC.

Jiang, R. & Murthy, D., 1996. *A mixture model involving three Weibull distributions.* In Proceedings of the Second Australia–Japan Workshop on Stochastic Models in Engineering, Technology and Management. Gold Coast, Australia, pp. 260-270.

Jiang, R. & Murthy, D., 1998. *Mixture of Weibull distributions - parametric characterization of failure rate function*. Applied Stochastic Models and Data Analysis, (14), 47-65.

Balakrishnan, N. & Rao, C.R., 2001. *Handbook of Statistics 20: Advances in Reliability 1st ed.*, Elsevier Science & Technology.

Relationship to Other Distributions

Weibull
Distribution

$Weibull(t; \alpha, \beta)$

Special Case:
$$Weibull(t; \alpha, \beta) = 2FWeibull(t; \alpha = \alpha_1, \beta = \beta_1, p = 1)$$
$$Weibull(t; \alpha, \beta) = 2FWeibull(t; \alpha = \alpha_2, \beta = \beta_2, p = 0)$$

3.2. Exponentiated Weibull Distribution

Probability Density Function - f(t)

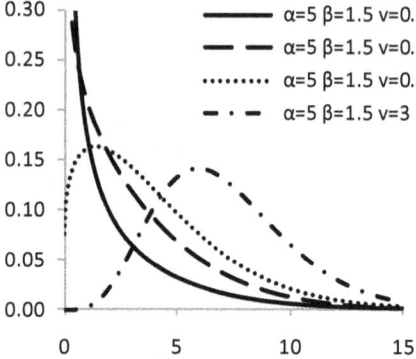

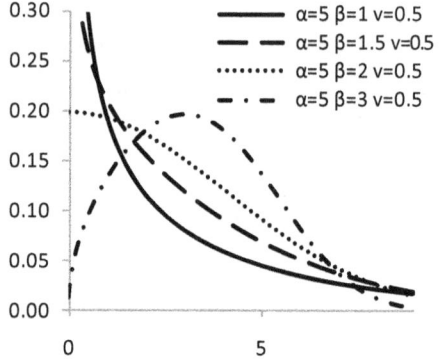

Cumulative Density Function - F(t)

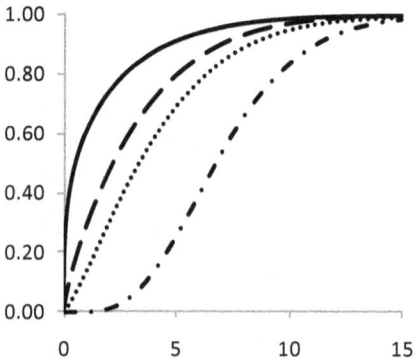

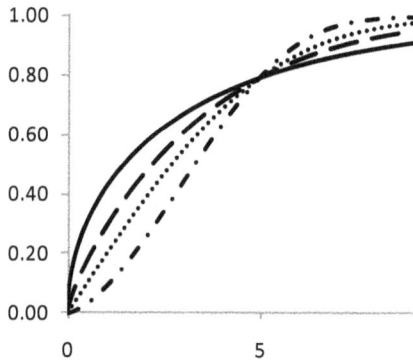

Hazard Rate - h(t)

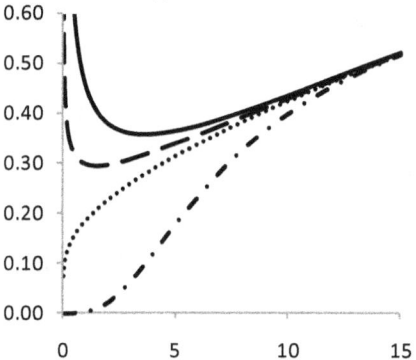

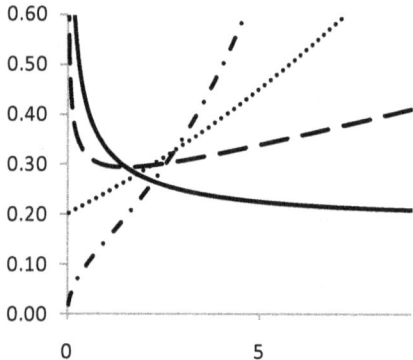

Parameters & Description		
Parameters	α $\alpha > 0$	Scale Parameter.
	β $\beta > 0$	Shape Parameter.
	v $v > 0$	Shape Parameter.
Limits	$t \geq 0$	

Distribution	Formulas
PDF	$$f(t) = \frac{\beta v t^{\beta-1}}{\alpha^{\beta}} \left[1 - \exp\left\{-\left(\frac{t}{\alpha}\right)^{\beta}\right\}\right]^{v-1} \exp\left\{-\left(\frac{t}{\alpha}\right)^{\beta}\right\}$$ $$= v\{F_W(t)\}^{v-1} f_W(t)$$ Where $F_W(t)$ and $f_W(t)$ are the cdf and pdf of the two parameter Weibull distribution, respectively.
CDF	$$F(t) = \left[1 - \exp\left\{-\left(\frac{t}{\alpha}\right)^{\beta}\right\}\right]^{v}$$ $$= [F_W(t)]^{v}$$
Reliability	$$R(t) = 1 - \left[1 - \exp\left\{-\left(\frac{t}{\alpha}\right)^{\beta}\right\}\right]^{v}$$ $$= 1 - [F_W(t)]^{v}$$
Conditional Survivor Function $P(T > x + t\|T > t)$	$$m(x) = R(x\|t) = \frac{R(t+x)}{R(t)} = \frac{1 - \left(1 - \exp\left\{-\left(\frac{t+x}{\alpha}\right)^{\beta}\right\}\right)^{v}}{1 - \left(1 - \exp\left\{-\left(\frac{t}{\alpha}\right)^{\beta}\right\}\right)^{v}}$$ Where t is the given time we know the component has survived to. x is a random variable defined as the time after t. Note: $x = 0$ at t.
Mean Residual Life	$$u(t) = \frac{\int_t^{\infty} \left[1 - \left(1 - \exp\left\{-\left(\frac{t}{\alpha}\right)^{\beta}\right\}\right)^{v}\right] dx}{1 - \left(1 - \exp\left\{-\left(\frac{t}{\alpha}\right)^{\beta}\right\}\right)^{v}}$$
Hazard Rate	$$h(t) = \frac{\beta v (t/\alpha)^{\beta-1} \left[1 - \exp\left\{-(t/\alpha)^{\beta}\right\}\right]^{v-1} \exp\left\{-(t/\alpha)^{\beta}\right\}}{1 - \left[1 - \exp\left\{-(t/\alpha)^{\beta}\right\}\right]^{v}}$$

For small t: (Murthy et al. 2003, p.130)

$$h(t) \approx \left(\frac{\beta v}{\alpha}\right)\left(\frac{t}{\alpha}\right)^{\beta v-1}$$

For large t: (Murthy et al. 2003, p.130)

$$h(t) \approx \left(\frac{\beta}{\alpha}\right)\left(\frac{t}{\alpha}\right)^{\beta-1}$$

Properties and Moments

Median

$$\alpha\left[-\ln\left\{1-2^{-1/v}\right\}\right]^{1/\beta}$$

Mode

For $\beta v > 1$ the mode can be approximated (Murthy, et al. 2003, p.130):

$$\alpha\left\{\frac{1}{2}\left[\frac{\sqrt{\beta(\beta-8v+2\beta v+9\beta v^2)}}{\beta v}-1-\frac{1}{v}\right]\right\}^{v}$$

Mean - 1st Raw Moment

Variance - 2nd Central Moment

Solved numerically see Murthy, et al. 2003, p.128

100p% Percentile Function

$$t_p = \alpha\left[-\ln\left(1-p^{1/v}\right)\right]^{1/\beta}$$

Parameter Estimation

Plotting Method (Jiang & Murthy 1999)

Plot Points on a Weibull Probability Plot	X-Axis	Y-Axis
	$x = ln(t)$	$y = ln\left[ln\left(\frac{1}{1-F}\right)\right]$

Using the Weibull Probability Plot the parameters can be estimated. (Jiang & Murthy 1999), provide a comprehensive coverage of this. A typical WPP for an exponentiated Weibull distribution is:

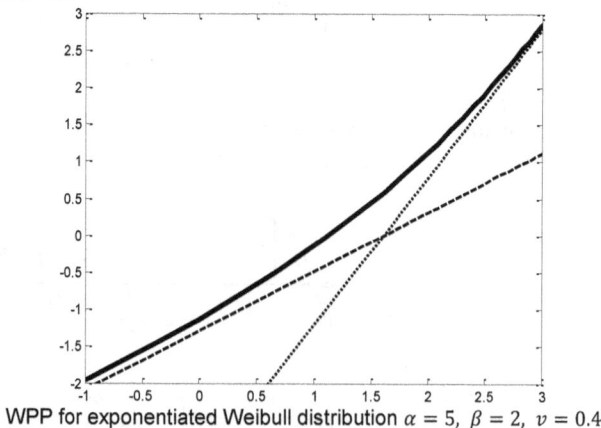

WPP for exponentiated Weibull distribution $\alpha = 5$, $\beta = 2$, $v = 0.4$

Asymptotes (Jiang & Murthy 1999):
As $x \rightarrow -\infty$ $(t \rightarrow 0)$ there exists an asymptote approximated by:

$$y \approx \beta v[x - \ln(\alpha)]$$

As $x \to \infty$ ($t \to \infty$) the asymptote straight line can be approximated by:
$$y \approx \beta[x - \ln(\alpha)]$$

Both asymptotes intersect the x-axis at $ln(\alpha)$ however both have different slopes unless $v = 1$ and the WPP is the same as a two parameter Weibull distribution.

Parameter Estimation
Plot estimates of the asymptotes ensuring they cross the x-axis at the same point. Use the right asymptote to estimate α and β. Use the left asymptote to estimate v.

Maximum Likelihood Bayesian	MLE and Bayesian techniques can be used in the standard way however estimates obtained from the graphical methods are useful for initial guesses when using numerical methods to solve equations. A literature review of MLE and Bayesian methods is covered in (Murthy et al. 2003).

Description, Limitations and Uses

Characteristics	**PDF Shape:** (Murthy et al. 2003, p.129) $\beta v <= 1$. The pdf is monotonically decreasing, $f(0) = \infty$. $\beta v = 1$. The pdf is monotonically decreasing, $f(0) = 1/\alpha$. $\beta v > 1$. The pdf is unimodal. $f(0) = 0$.

The pdf shape is determined by βv in a similar way to the β for a two parameter Weibull distribution.

Hazard Rate Shape: (Murthy et al. 2003, p.129)
$\beta \le 1$ and $\beta v \le 1$. The hazard rate is monotonically decreasing.
$\beta \ge 1$ and $\beta v \ge 1$. The hazard rate is monotonically increasing.
$\beta < 1$ and $\beta v > 1$. The hazard rate is unimodal.
$\beta > 1$ and $\beta v < 1$. The hazard rate is a bathtub curve.

Weibull Distribution. The Weibull distribution is a special case of the expatiated distribution when $v = 1$. When v is an integer greater than 1, then the cdf represents a multiplicative Weibull model.

Standard Exponentiated Weibull. (Xie et al. 2004) When $\alpha = 1$ the distribution is the standard exponentiated Weibull distribution with cdf:
$$F(t) = \left[1 - \exp\{-t^\beta\}\right]^v$$

Minimum Failure Rate. (Xie et al. 2004) When the hazard rate is a bathtub curve ($\beta > 1$ and $\beta v < 1$) then the minimum failure rate point is:
$$t' = \alpha[-\ln(1 - y_1)]^{1/\beta}$$

where y_1 is the solution to:
$$(\beta - 1)y(1 - y^v) + \beta \ln(1 - y)[1 + vy - v - y^v] = 0$$

Maximum Mean Residual Life. (Xie et al. 2004) By solving the derivative of the MRL function to zero, the maximum MRL is found by solving to t:

$$t^* = \alpha[-\ln(1 - y_2)]^{1/\beta}$$

where y_2 is the solution to:

$$\beta v(1 - y)y^{v-1}[-\ln(1 - y)]^{-1/\beta}$$
$$\times \int_{[-\ln(1-y)]^{1/\beta}}^{\infty} [1 - (1 - e^{-x^\beta})^v] \, dx - (1 - y^v)^2 = 0$$

Resources

Books / Journals:
Mudholkar, G. & Srivastava, D., 1993. *Exponentiated Weibull family for analyzing bathtub failure-rate data*. Reliability, IEEE Transactions on, 42(2), 299-302.

Jiang, R. & Murthy, D., 1999. *The exponentiated Weibull family: a graphical approach. Reliability*, IEEE Transactions on, 48(1), 68-72.

Xie, M., Goh, T.N. & Tang, Y., 2004. *On changing points of mean residual life and failure rate function for some generalized Weibull distributions.* Reliability Engineering and System Safety, 84(3), 293–299.

Murthy, D., Xie, M. & Jiang, R., 2003. *Weibull Models* 1st ed., Wiley-Interscience.

Rinne, H., 2008. The *Weibull Distribution: A Handbook* 1st ed., Chapman & Hall/CRC.

Balakrishnan, N. & Rao, C.R., 2001. *Handbook of Statistics 20: Advances in Reliability 1st ed.*, Elsevier Science & Technology.

Relationship to Other Distributions

Weibull Distribution

Special Case:
$$Weibull(t; \alpha, \beta) = ExpWeibull(t; \alpha = \alpha, \beta = \beta, v = 1)$$

$Weibull(t; \alpha, \beta)$

3.3. Modified Weibull Distribution

Probability Density Function - f(t)

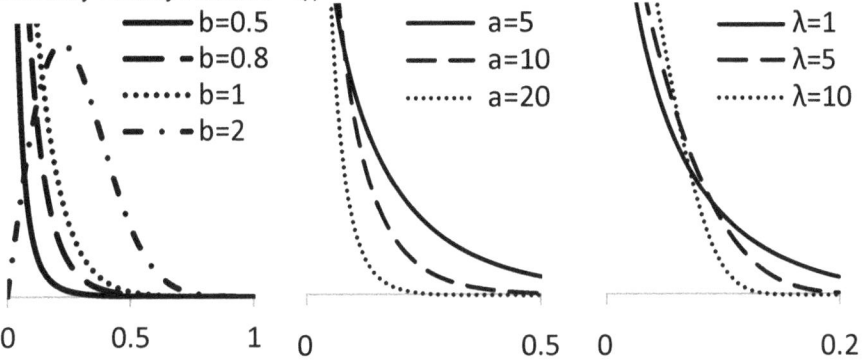

Cumulative Density Function - F(t)

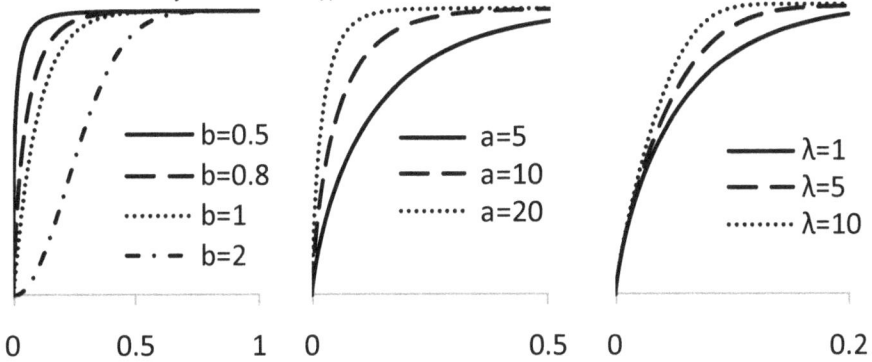

Hazard Rate - h(t)

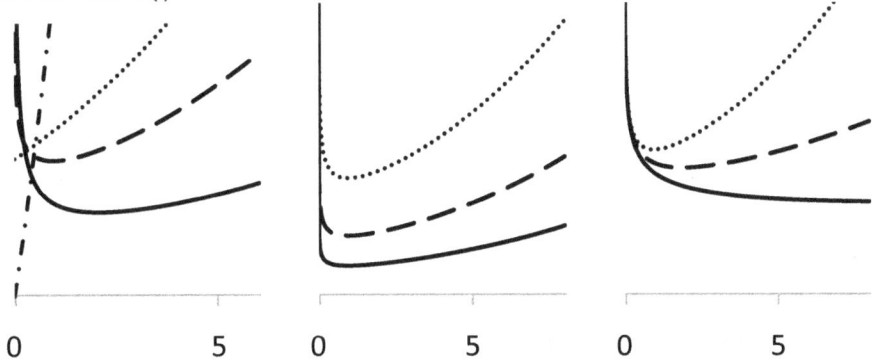

Note: The hazard rate plots are on a different scale to the PDF and CDF

Parameters & Description

Parameters			
	a	$a > 0$	*Scale Parameter.*
	b	$b \geq 0$	*Shape Parameter:* The shape of the distribution is completely determined by b. When $0 < b < 1$ the distribution has a bathtub shaped hazard rate.
	λ	$\lambda \geq 0$	*Scale Parameter.*
Limits		$t \geq 0$	

Distribution	Formulas
PDF	$f(t) = a(b + \lambda t)\, t^{b-1} \exp(\lambda t)\, \exp\left[-at^b \exp(\lambda t)\right]$
CDF	$F(t) = 1 - \exp[-at^b\, exp(\lambda t)]$
Reliability	$R(t) = \exp[-at^b\, exp(\lambda t)]$
Mean Residual Life	$u(t) = \exp\left(at^b e^{\lambda t}\right) \int_t^\infty \exp\left(ax^b e^{\lambda t}\right) dx$
Hazard Rate	$h(t) = a(b + \lambda t)t^{b-1}e^{\lambda t}$

Properties and Moments

Median	Solved numerically (see 100p%)
Mode	Solved numerically
Mean - 1st Raw Moment	Solved numerically
Variance - 2nd Central Moment	Solved numerically
100p% Percentile Function	Solve for t_p numerically:

$$t_p^b \exp(\lambda t_p) = -\frac{\ln(1 - p)}{a}$$

Parameter Estimation

Plotting Method (Lai et al. 2003)

Plot Points on a Weibull Probability Plot	X-Axis	Y-Axis
	$\ln(t_i)$	$ln\left[ln\left(\dfrac{1}{1-F}\right)\right]$

Using the Weibull Probability Plot the parameters can be estimated. (Lai et al. 2003).

Asymptotes (Lai et al. 2003):
As $x \to -\infty$ $(t \to 0)$ the asymptote straight line can be approximated as:
$$y \approx bx + \ln(a)$$

As $x \to \infty$ $(t \to \infty)$ the asymptote straight line can be approximated as (not used for parameter estimate but more for model validity):
$$y \approx \lambda \exp(x) = \lambda t$$

Intersections (Lai et al. 2003):

Y-Axis Intersection $(0, x_0)$
$$\ln(a) + bx_0 + \lambda e^{x_0} = 0$$
X-Axis Intersection $(y_0, 0)$
$$\ln(a) + \lambda = y_0$$

Solving these gives an approximate value for each parameter which can be used as an initial guess for numerical methods solving MLE or Bayesian methods.

A typical WPP for a Modified Weibull Distribution is:

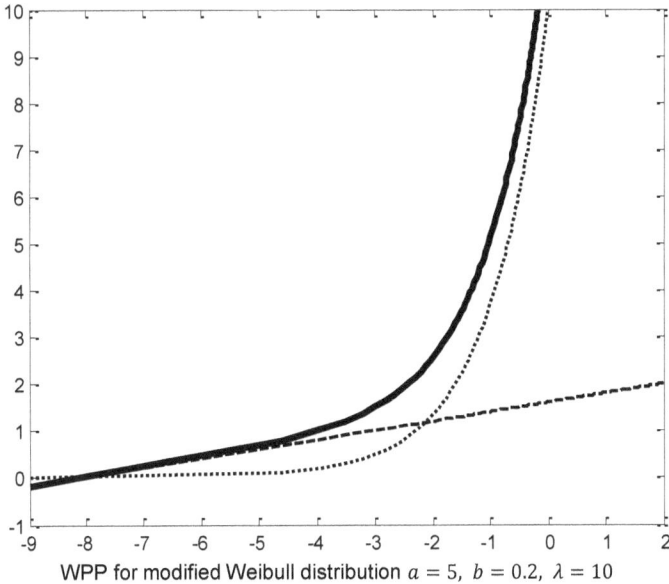

WPP for modified Weibull distribution $a = 5$, $b = 0.2$, $\lambda = 10$

Description, Limitations and Uses

Characteristics   **Parameter Characteristics:**(Lai et al. 2003)
$0 < b < 1$ and $\lambda > 0$. The hazard rate has a bathtub curve shape. $h(t) \to \infty$ as $t \to 0$. $h(t) \to \infty$ as $t \to \infty$.
$b \geq 1$ and $\lambda > 0$, has an increasing hazard rate function. $h(0) = 0$. $h(t) \to \infty$ as $t \to \infty$.
$\lambda = 0$. The function has the same form as a Weibull Distribution. $h(0) = ab$. $h(t) \to \infty$ as $t \to \infty$
Minimum Failure Rate. (Xie et al. 2004) When the hazard rate is a bathtub curve $(0 < b < 1$ and $\lambda > 0)$ then the minimum failure rate point is given as:

$$t^* = \frac{\sqrt{b} - b}{\lambda}$$

Maximum Mean Residual Life. (Xie et al. 2004) By solving the derivative of the MRL function to zero, the maximum MRL is found by solving to t:

$$a(b + \lambda t)t^{b-1}e^{\lambda t}\int_t^\infty \exp(-ax^b e^{(\lambda x)dx} - \exp(at^b e^{\lambda t}) = 0$$

Shape. The shape of the hazard rate cannot have a flat "usage period" and a strong "wear out" gradient.

Resources

Books / Journals:
Lai, C., Xie, M. & Murthy, D., 2003. *A modified Weibull distribution*. IEEE Transactions on Reliability, 52(1), 33-37.

Murthy, D.N.P., Xie, M. & Jiang, R., 2003. *Weibull Models* 1st ed., Wiley-Interscience.

Xie, M., Goh, T.N. & Tang, Y., 2004. *On changing points of mean residual life and failure rate function for some generalized Weibull distributions*. Reliability Engineering and System Safety, 84(3), 293–299.

Rinne, H., 2008. The *Weibull Distribution: A Handbook* 1st ed., Chapman & Hall/CRC.

Balakrishnan, N. & Rao, C.R., 2001. *Handbook of Statistics 20: Advances in Reliability* 1st ed., Elsevier Science & Technology.

Relationship to Other Distributions

Weibull Distribution

$Weibull(t; \alpha, \beta)$

Special Case:
$$Weibull(t; \alpha, \beta) = ModWeibull(t; a = \alpha, b = \beta, \lambda = 0)$$

4. Univariate Continuous Distributions

4.1. Beta Continuous Distribution

Probability Density Function - f(t)

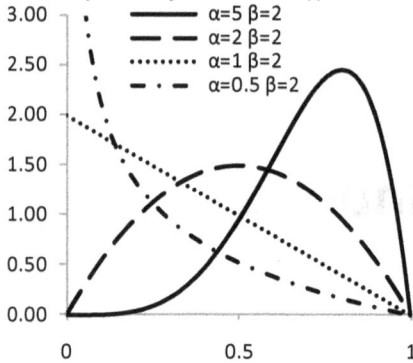

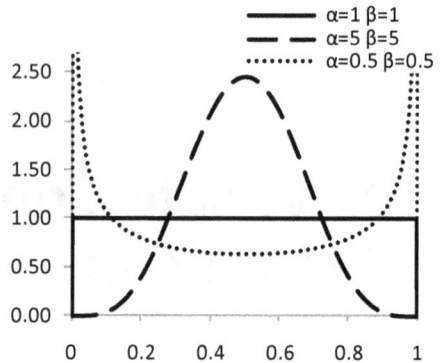

Cumulative Density Function - F(t)

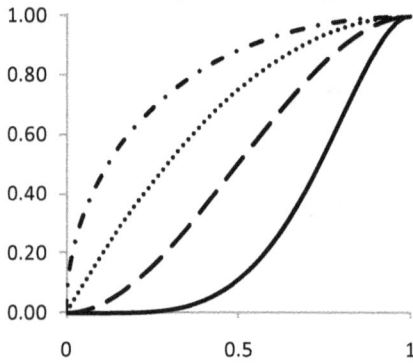

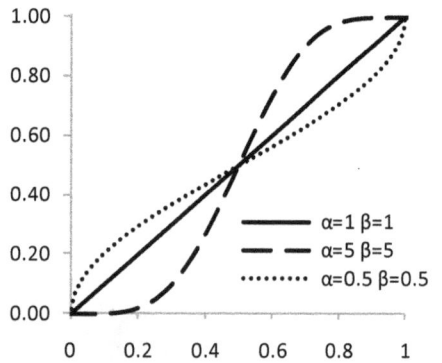

Hazard Rate - h(t)

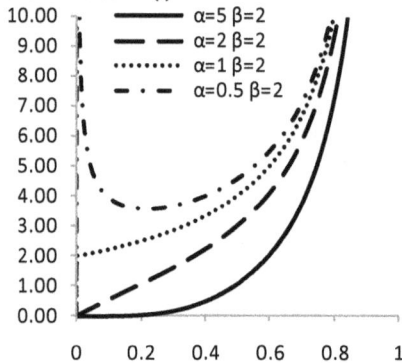

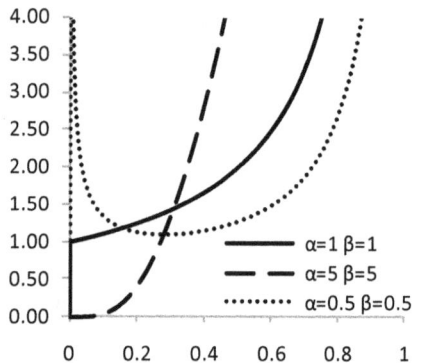

Parameters & Description

Parameters			
	α	$\alpha > 0$	*Shape Parameter.*
	β	$\beta > 0$	*Shape Parameter.*
	a_L	$-\infty < a_L < b_U$	*Lower Bound:* a_L is the lower bound but has also been called a location parameter. In the standard Beta distribution $a_L = 0$.
	b_U	$a_L < b_U < \infty$	*Upper Bound:* b_U is the upper bound. In the standard Beta distribution $b_U = 1$. The scale parameter may also be defined as $b_U - a_L$.
Limits		$a_L < t \leq b_U$	

Distribution	Formulas

$B(x,y)$ is the Beta function, $B_t(t|x,y)$ is the incomplete Beta function, $I_t(t|x,y)$ is the regularized Beta function, $\Gamma(k)$ is the complete gamma which is discussed in Section 1.6.

PDF

General Form:

$$f(t; \alpha, \beta, a_L, b_U) = \frac{\Gamma(\alpha + \beta)}{\Gamma(\alpha)\Gamma(\beta)} \cdot \frac{(t - a_L)^{\alpha-1}(b_U - t)^{\beta-1}}{(b_U - a_L)^{\alpha+\beta-1}}$$

When $a_L = 0, b_U = 1$:

$$f(t|\alpha, \beta) = \frac{\Gamma(\alpha + \beta)}{\Gamma(\alpha)\Gamma(\beta)} \cdot t^{\alpha-1}(1 - t)^{\beta-1}$$

$$= \frac{1}{B(\alpha, \beta)} \cdot t^{\alpha-1}(1 - t)^{\beta-1}$$

CDF

$$F(t) = \frac{\Gamma(\alpha + \beta)}{\Gamma(\alpha)\Gamma(\beta)} \int_0^t u^{\alpha-1}(1 - u)^{\beta-1} \, du$$

$$= \frac{B_t(t|\alpha, \beta)}{B(\alpha, \beta)}$$

$$= I_t(t|\alpha, \beta)$$

Reliability

$$R(t) = 1 - I_t(t|\alpha, \beta)$$

Conditional Survivor Function

$$m(x) = R(x|t) = \frac{R(t + x)}{R(t)} = \frac{1 - I_t(t + x|\alpha, \beta)}{1 - I_t(t|\alpha, \beta)}$$

Where
t is the given time we know the component has survived to.
x is a random variable defined as the time after t. Note: $x = 0$ at t.

Mean Residual Life

$$u(t) = \frac{\int_t^\infty \{B(\alpha, \beta) - B_x(x|\alpha, \beta)\}dx}{B(\alpha, \beta) - B_t(t|\alpha, \beta)}$$

(Gupta and Nadarajah 2004, p.44)

Hazard Rate

$$h(t) = \frac{t^{\alpha-1}(1 - t)}{B(\alpha, \beta) - B_t(t|\alpha, \beta)}$$

(Gupta and Nadarajah 2004, p.44)

Properties and Moments

Median

Numerically solve for t:
$$t_{0.5} = F^{-1}(\alpha, \beta)$$

Mode

$$\frac{\alpha-1}{\alpha+\beta-2} \text{ for } \alpha > 1 \text{ and } \beta > 1$$

Mean - 1st Raw Moment

$$\frac{\alpha}{\alpha + \beta}$$

Variance - 2nd Central Moment

$$\frac{\alpha\beta}{(\alpha + \beta)^2(\alpha + \beta + 1)}$$

Skewness - 3rd Central Moment

$$\frac{2(\beta - \alpha)\sqrt{\alpha + \beta + 1}}{(\alpha + \beta + 2)\sqrt{\alpha\beta}}$$

Excess kurtosis - 4th Central Moment

$$\frac{6[\alpha^3 + \alpha^2(1 - 2\beta) + \beta^2(1 + \beta) - 2\alpha\beta(2 + \beta)]}{\alpha\beta(\alpha + \beta + 2)(\alpha + \beta + 3)}$$

Characteristic Function

$$_1F_1(\alpha; \alpha + \beta; it)$$

Where $_1F_1$ is the confluent hypergeometric function defined as:

$$_1F_1(\alpha; \beta; x) = \sum_{k=0}^{\infty} \frac{(\alpha)_k}{(\beta)_k} \cdot \frac{x^k}{k!}$$

(Gupta and Nadarajah 2004, p.44)

100p% Percentile Function

Numerically solve for t:
$$t_p = F^{-1}(\alpha, \beta)$$

Parameter Estimation

Maximum Likelihood Function

Likelihood Functions

$$L(\alpha, \beta|E) = \underbrace{\frac{\Gamma(\alpha + \beta)n_F}{\Gamma(\alpha)\Gamma(\beta)} \prod_{i=1}^{n_F} t_i^{F\ \alpha-1}\left(1 - t_i^F\right)^{\beta-1}}_{failures}.$$

Log-Likelihood Functions

$$\Lambda(\alpha, \beta|E) = n_F\{\ln[\Gamma(\alpha + \beta) - ln[\Gamma(\alpha)] - ln[\Gamma(\beta)]\}$$
$$+(\alpha - 1)\sum_{i=1}^{n_F} \ln(t_i^F) + (\beta - 1)\sum_{i=1}^{n_F} \ln(1 - t_i^F)$$

$$\frac{\partial\Lambda}{\partial\alpha} = 0$$

$$\psi(\alpha) - \psi(\alpha + \beta) = \frac{1}{n_F}\sum_{i=1}^{n_F} \ln(t_i^F)$$

where $\psi(x) = \frac{d}{dx}\ln[\Gamma(x)]$ is the digamma function see Section1.6.7.
(Johnson et al. 1995, p.223)

$$\frac{\partial\Lambda}{\partial\beta} = 0$$

$$\psi(\beta) - \psi(\alpha + \beta) = \frac{1}{n_F}\sum_{i=1}^{n_F} \ln(1 - t_i)$$

(Johnson et al. 1995, p.223)

| Point Estimates | Point estimates are obtained by using numerical methods to solve the simultaneous equations above. |

Fisher Information Matrix

$$I(\alpha, \beta) = \begin{bmatrix} \psi'(\alpha) - \psi'(\alpha + \beta) & -\psi'(\alpha + \beta) \\ -\psi'(\alpha + \beta) & \psi'(\beta) - \psi'(\alpha + \beta) \end{bmatrix}$$

where $\psi'(x) = \frac{d^2}{dx^2} \ln\Gamma(x) = \sum_{i=0}^{\infty}(x + i)^{-2}$ is the Trigamma function. See Section 1.6.8. (Yang and Berger 1998, p.5)

Confidence Intervals

For large number of samples, the Fisher information matrix can be used to estimate confidence intervals. See Section 1.4.7.

Bayesian

Non-informative Priors

Jeffery's Prior

$$\sqrt{\det(I(\alpha, \beta))}$$

where $I(\alpha, \beta)$ is given above.

Conjugate Priors

UOI	Likelihood Model	Evidence	Dist. of UOI	Prior Para	Posterior Parameters
p from $Bernoulli(k; p)$	Bernoulli	k failures in 1 trail	Beta	α_0, β_0	$\alpha = \alpha_o + k$ $\beta = \beta_o + 1 - k$
p from $Binom(k; p, n)$	Binomial	k failures in n trials	Beta	α_0, β_0	$\alpha = \alpha_o + k$ $\beta = \beta_o + n - k$

Description, Limitations and Uses

| Example | For examples on the use of the beta distribution as a conjugate prior see the binomial distribution. |

A non-homogeneous (operate in different environments) population of 5 switches have the following probabilities of failure on demand.

0.1176,   0.1488,   0.3684,   0.8123,   0.9783

Estimate the population variability function:

$$\frac{1}{n_F} \sum_{i=1}^{n_F} \ln(t_i^F) = -1.0549$$

$$\frac{1}{n_F} \sum_{i=1}^{n_F} \ln(1 - t_i) = -1.25$$

Numerically Solving:

$$\psi(\alpha) + 1.0549 = \psi(\beta) + 1.25$$

Gives:

$$\hat{\alpha} = 0.7369$$
$$\hat{b} = 0.6678$$

$$I(\alpha,\beta) = \begin{bmatrix} 1.5924 & -1.0207 \\ -1.0207 & 2.0347 \end{bmatrix}$$

$$[J_n(\hat{\alpha},\hat{\beta})]^{-1} = [n_F I(\hat{\alpha},\hat{\beta})]^{-1} = \begin{bmatrix} 0.1851 & 0.0929 \\ 0.0929 & 0.1449 \end{bmatrix}$$

90% confidence interval for α:

$$\left[\hat{\alpha}.\exp\left\{ \frac{\Phi^{-1}(0.95)\sqrt{0.1851}}{-\hat{\alpha}} \right\}, \quad \hat{\alpha}.\exp\left\{ \frac{\Phi^{-1}(0.95)\sqrt{0.1851}}{\hat{\alpha}} \right\} \right]$$

$$[0.282, \quad 1.92]$$

90% confidence interval for β:

$$\left[\hat{\beta}.\exp\left\{ \frac{\Phi^{-1}(0.95)\sqrt{0.1449}}{-\hat{\beta}} \right\}, \quad \hat{\beta}.\exp\left\{ \frac{\Phi^{-1}(0.95)\sqrt{0.1449}}{\hat{\beta}} \right\} \right]$$

$$[0.262, \quad 1.71]$$

Characteristics

The Beta distribution was originally known as a Pearson Type I distribution (and Type II distribution which is a special case of a Type I).

$Beta(\alpha,\beta)$ is the mirror distribution of $Beta(\beta,\alpha)$. If $X \sim Beta(\alpha,\beta)$ and let $Y = 1 - X$ then $Y \sim Beta(\beta,\alpha)$.

Location / Scale Parameters (NIST Section 1.3.6.6.17)
a_L and b_U can be transformed into a location and scale parameter:
$$location = a_L$$
$$scale = b_U - a_L$$

Shapes (Gupta and Nadarajah 2004, p.41):
$0 < \alpha < 1$. As $x \to 0, f(x) \to \infty$.
$0 < \beta < 1$. As $x \to 1, f(x) \to \infty$.
$\alpha > 1$, $\beta > 1$. As $x \to 0, f(x) \to 0$. There is a single mode at $\frac{\alpha-1}{\alpha+\beta-2}$.
$\alpha < 1$, $\beta < 1$. The distribution is a U shape. There is a single anti-mode at $\frac{\alpha-1}{\alpha+\beta-2}$.
$\alpha > 0$, $\beta > 0$. There exist inflection points at:
$$\frac{\alpha-1}{\alpha+\beta-2} \pm \frac{1}{\alpha+\beta-2}.\sqrt{\frac{(\alpha-1)(\beta-1)}{\alpha+\beta-3}}$$
$\alpha = \beta$. The distribution is symmetrical about $x = 0.5$. As $\alpha = \beta$ becomes large, the beta distribution approaches the normal distribution. The Standard Uniform Distribution arises when $\alpha = \beta = 1$.
$\alpha = 1$, $\beta = 2$ or $\alpha = 2$, $\beta = 1$. Straight line.
$(\alpha-1)(\beta-1) < 0$. J Shaped.

Hazard Rate and MRL (Gupta and Nadarajah 2004, p.45):
$\alpha \geq 1$, $\beta \geq 1$. $h(t)$ is increasing. $u(t)$ is decreasing.

$\alpha \leq 1$, $\beta \leq 1$. $h(t)$ is decreasing. $u(t)$ is increasing.
$\alpha > 1$, $0 < \beta < 1$. $h(t)$ is bathtub shaped and $u(t)$ is an upside-down bathtub shape.
$0 < \alpha < 1$, $\beta > 1$. $h(t)$ is upside down bathtub shaped and $u(t)$ is bathtub shape.

Parameter Model. The Beta distribution is often used to model parameters which are constrained to take place between an interval. In particular the distribution of a probability parameter $0 \leq p \leq 1$ is popular with the Beta distribution.

Applications

Bayesian Analysis. The Beta distribution is often used as a conjugate prior in Bayesian analysis for the Bernoulli, Binomial and Geometric Distributions to produce closed form posteriors. The $Beta(0,0)$ distribution is an improper prior, sometimes used to represent ignorance of parameter values. The $Beta(1,1)$ is a standard uniform distribution which may be used as a non-informative prior. When used as a conjugate prior to a Bernoulli or Binomial process the parameter α may represent the number of successes and β the total number of failures with the total number of trials being $n = \alpha + \beta$.

Proportions. Used to model proportions. An example of this is the likelihood ratios for estimating uncertainty.

Online:
http://mathworld.wolfram.com/BetaDistribution.html
http://en.wikipedia.org/wiki/Beta_distribution
http://socr.ucla.edu/htmls/SOCR_Distributions.html (interactive web calculator)
http://www.itl.nist.gov/div898/handbook/eda/section3/eda366h.htm

Resources

Books:
Gupta, A.K. & Nadarajah, S., 2004. *Handbook of beta distribution and its applications*, CRC Press.

Johnson, N.L., Kotz, S. & Balakrishnan, N., 1995. *Continuous Univariate Distributions*, Vol. 2 2nd ed., Wiley-Interscience.

Relationship to Other Distributions

Chi-square Distribution

$\chi^2(t; v)$

Let

$$X_i \sim \chi^2(v_i) \qquad and \qquad Y = \frac{X_1}{X_1 + X_2}$$

Then

$$Y \sim Beta\left(\alpha = \tfrac{1}{2}v_1, \beta = \tfrac{1}{2}v_2\right)$$

Uniform Distribution

Let

$$X_i \sim Unif(0,1) \qquad and \qquad X_1 \leq X_2 \leq \cdots \leq X_n$$

Then

$Unif(t; a, b)$

$$X_r \sim Beta(r, n - r + 1)$$
Where n and r are integers.

Special Case:
$$Beta(t; 1,1, a, b) = Unif(t; a, b)$$

For large α and β with fixed α/β:

Normal Distribution

$$Beta(\alpha, \beta) \approx Norm\left(\mu = \frac{\alpha}{\alpha + \beta}, \sigma = \sqrt{\frac{\alpha\beta}{(\alpha + \beta)^2(\alpha + \beta + 1)}}\right)$$

$Norm(t; \mu, \sigma)$

As α and β increase the mean remains constant and the variance is reduced.

Gamma Distribution

Let
$$X_1, X_2 \sim Gamma(k_i, \lambda_i) \quad and \quad Y = \frac{X_1}{X_1 + X_2}$$

$Gamma(t; k, \lambda)$

Then
$$Y \sim Beta(\alpha = k_1, \beta = k_2)$$

Dirichlet Distribution

Special Case:
$$Dir_{d=1}(x; [\alpha_1, \alpha_0]) = Beta(k = x; \alpha = \alpha_1, \beta = \alpha_0)$$

$Dir_d(x; \boldsymbol{\alpha})$

4.2. Birnbaum Saunders Continuous Distribution

Probability Density Function - f(t)

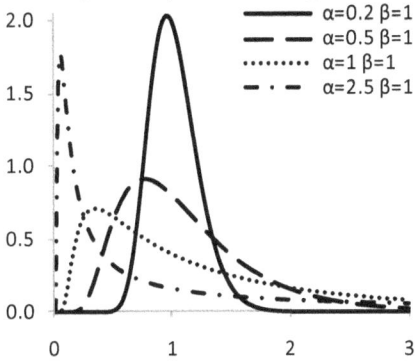

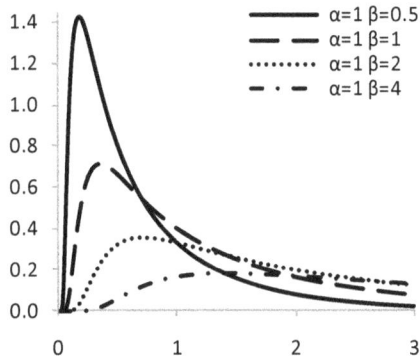

Cumulative Density Function - F(t)

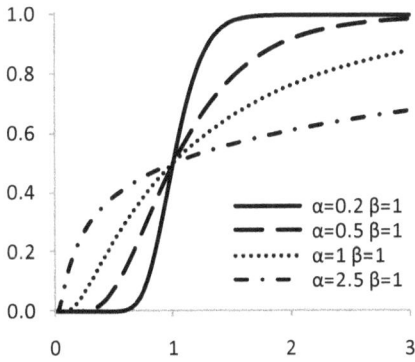

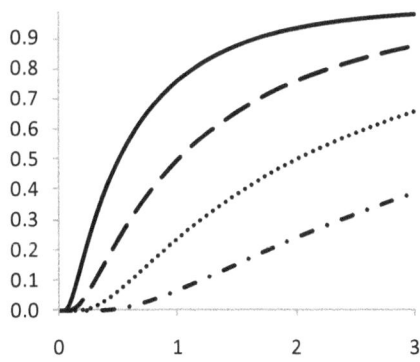

Hazard Rate - h(t)

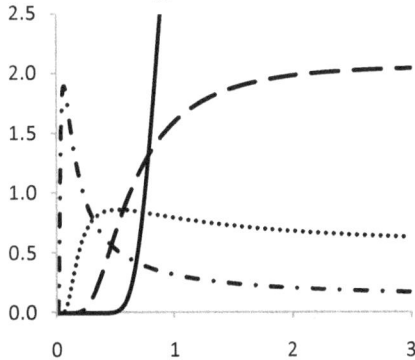

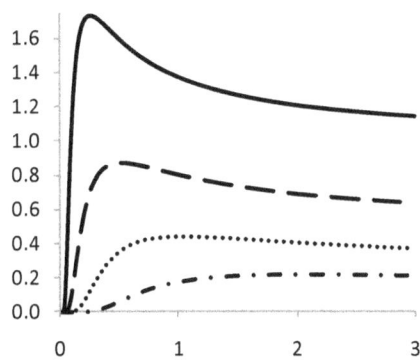

Parameters & Description

Parameters	β	$\beta > 0$	Scale parameter. β is the scale parameter equals to the median.
	α	$\alpha > 0$	Shape parameter.
Limits		$0 < t < \infty$	

Distribution — Formulas

PDF

$$f(t) = \frac{\sqrt{t/\beta} + \sqrt{\beta/t}}{2\alpha t \sqrt{2\pi}} \exp\left[-\frac{1}{2}\left(\frac{\sqrt{t/\beta} - \sqrt{\beta/t}}{\alpha} \right)^2 \right]$$

$$= \frac{\sqrt{t/\beta} + \sqrt{\beta/t}}{2\alpha t} \phi(z)$$

where $\phi(z)$ is the standard normal pdf and:

$$z_{BS} = \frac{\sqrt{t/\beta} - \sqrt{\beta/t}}{\alpha}$$

CDF

$$F(t) = \Phi\left(\frac{\sqrt{t/\beta} - \sqrt{\beta/t}}{\alpha} \right)$$

$$= \Phi(z_{BS})$$

Reliability

$$R(t) = \Phi\left(\frac{\sqrt{\beta/t} - \sqrt{t/\beta}}{\alpha} \right)$$

$$= \Phi(-z_{BS})$$

Conditional Survivor Function $P(T > x + t | T > t)$

$$m(x) = R(x|t) = \frac{R(t+x)}{R(t)} = \frac{\Phi(-z'_{BS})}{\Phi(-z_{BS})}$$

Where

$$z_{BS} = \frac{\sqrt{t/\beta} - \sqrt{\beta/t}}{\alpha}, \qquad z'_{BS} = \frac{\sqrt{(t+x)/\beta} - \sqrt{\beta/(t+x)}}{\alpha}$$

t is the given time we know the component has survived to.
x is a random variable defined as the time after t. Note: $x = 0$ at t.

Mean Residual Life

$$u(t) = \frac{\int_t^\infty \Phi(-z_{BS}) dx}{\Phi(-z_{BS})}$$

Hazard Rate

$$h(t) = \frac{\sqrt{t/\beta} + \sqrt{\beta/t}}{2\alpha t} \left[\frac{\phi(z_{BS})}{\Phi(-z_{BS})} \right]$$

Cumulative Hazard Rate

$$H(t) = -\ln[\Phi(-z_{BS})]$$

Properties and Moments

Median	β
Mode	Numerically solve for t: $$t^3 + \beta(1+\alpha^2)t^2 + \beta^2(3\alpha^2 - 1)t - \beta^3 = 0$$
Mean - 1st Raw Moment	$$\beta\left(1 + \frac{\alpha^2}{2}\right)$$
Variance - 2nd Central Moment	$$\alpha^2\beta^2\left(1 + \frac{5\alpha^2}{4}\right)$$
Skewness - 3rd Central Moment	$$\frac{4\alpha(11\alpha^2 + 6)}{(5\alpha^2 + 4)^{\frac{3}{2}}}$$ (Lemonte et al. 2007)
Excess kurtosis - 4th Central Moment	$$3 + \frac{6\alpha^2(93\alpha^2 + 40)}{(5\alpha^2 + 4)^2}$$ (Lemonte et al. 2007)
100γ % Percentile Function	$$t_\gamma = \frac{\beta}{4}\left\{\alpha\Phi^{-1}(\gamma) + \sqrt{4 + [\alpha\Phi^{-1}(\gamma)]^2}\right\}^2$$

Parameter Estimation

Maximum Likelihood Function

Likelihood Function	For complete data: $$L(\theta, \alpha|E) = \underbrace{\prod_{i=1}^{n_F} \frac{\sqrt{t_i/\beta} + \sqrt{\beta/t_i}}{2\alpha t_i \sqrt{2\pi}} \exp\left[-\frac{1}{2}\left(\frac{\sqrt{t_i/\beta} - \sqrt{\beta/t_i}}{\alpha}\right)^2\right]}_{\text{failures}}$$
Log-Likelihood Function	$$\Lambda(\alpha, \beta|E) = \underbrace{-n_F \ln(\alpha\beta) + \sum_{i=1}^{n_F} \ln\left[\left(\frac{\beta}{t_i}\right)^{\frac{1}{2}} + \left(\frac{\beta}{t_i}\right)^{\frac{3}{2}}\right] - \frac{1}{2\alpha^2}\sum_{i=1}^{n_F}\left(\frac{t_i}{\beta} + \frac{\beta}{t_i} - 2\right)}_{\text{failures}}$$
$\dfrac{\partial\Lambda}{\partial\alpha} = 0$	$$\underbrace{\frac{\partial\Lambda}{\partial\alpha} = -\frac{n_F}{\alpha}\left(1 + \frac{2}{\alpha^2}\right) + \frac{1}{\alpha^3\beta}\sum_{i=1}^{n_F} t_i + \frac{\beta}{\alpha^3}\sum_{i=1}^{n_F}\frac{1}{t_i}}_{\text{failures}} = 0$$
$\dfrac{\partial\Lambda}{\partial\beta} = 0$	$$\underbrace{\frac{\partial\Lambda}{\partial\beta} = -\frac{n_F}{2\beta} + \sum_{i=1}^{n_F}\frac{1}{t_i + \beta} + \frac{1}{2\alpha^2\beta^2}\sum_{i=1}^{n_F} t_i - \frac{1}{2\alpha^2}\sum_{i=1}^{n_F}\frac{1}{t_i}}_{\text{failures}} = 0$$
MLE Point Estimates	$\hat{\beta}$ is found by solving:

$$\beta^2 - \beta[2R + g(\beta)] + R[S + g(\beta)] = 0$$

where

$$g(\beta) = \left[\frac{1}{n}\sum_{i=1}^{n_F}\frac{1}{\beta + t_i}\right]^{-1}, \qquad S = \frac{1}{n_F}\sum_{i=1}^{n_F}t_i, \qquad R = \left(\frac{1}{n_F}\sum_{i=1}^{n_F}\frac{1}{t_i}\right)^{-1}$$

Point estimates for $\hat{\alpha}$ is:

$$\hat{\alpha} = \sqrt{\frac{S}{\hat{\beta}} + \frac{\hat{\beta}}{R} - 2}$$

(Lemonte et al. 2007)

Fisher Information

$$I(\theta, \alpha) = \begin{bmatrix} \dfrac{2}{\alpha^2} & 0 \\ 0 & \dfrac{\alpha(2\pi)^{-1/2}k(\alpha) + 1}{\alpha^2\beta^2} \end{bmatrix}$$

where

$$k(\alpha) = \alpha\sqrt{\frac{\pi}{2}} - \pi\exp\left\{\frac{2}{\alpha^2}\right\}\left[1 - \Phi\left(\frac{2}{\alpha}\right)\right]$$

(Lemonte et al. 2007)

100γ% Confidence Intervals

Calculated from the Fisher information matrix. See section 1.4.7. For a literature review of proposed confidence intervals see (Lemonte et al. 2007).

Description, Limitations and Uses

Example

5 components are put on a test with the following failure times: 98, 116, 2485, 2526, 2920 hours

$$S = \frac{1}{n_F}\sum_{i=1}^{n_F}t_i = 1629$$

$$R = \left(\frac{1}{n_F}\sum_{i=1}^{n_F}\frac{1}{t_i}\right)^{-1} = 250.432$$

Solving:

$$\beta^2 - \beta\left\{2R + \left[\frac{1}{n}\sum_{i=1}^{n_F}\frac{1}{\beta + t_i}\right]^{-1}\right\} + R\left\{S + \left[\frac{1}{n}\sum_{i=1}^{n_F}\frac{1}{\beta + t_i}\right]^{-1}\right\} = 0$$

$$\hat{\beta} = 601.949$$

$$\hat{\alpha} = \sqrt{\frac{S}{\hat{\beta}} + \frac{\hat{\beta}}{R} - 2} = 1.763$$

90% confidence interval for α:

$$\left[\hat{\alpha}.\exp\left\{\frac{\Phi^{-1}(0.95)\sqrt{\frac{\alpha^2}{2n_F}}}{-\hat{\alpha}}\right\}, \quad \hat{\alpha}.\exp\left\{\frac{\Phi^{-1}(0.95)\sqrt{\frac{\alpha^2}{2n_F}}}{\hat{\alpha}}\right\}\right]$$

$$[1.048, \quad 2.966]$$

90% confidence interval for β:

$$k(\hat{\alpha}) = \hat{\alpha}\sqrt{\frac{\pi}{2}} - \pi\exp\left\{\frac{2}{\hat{\alpha}^2}\right\}\left[1 - \Phi\left(\frac{2}{\hat{\alpha}}\right)\right] = 1.442$$

$$I_{\beta\beta} = \frac{\hat{\alpha}(2\pi)^{-1/2}k(\hat{\alpha}) + 1}{\hat{\alpha}^2\hat{\beta}^2} = 10.34E\text{-}6$$

$$\left[\hat{\beta}.\exp\left\{\frac{\Phi^{-1}(0.95)\sqrt{\frac{96762}{n_F}}}{-\hat{\beta}}\right\}, \quad \hat{\beta}.\exp\left\{\frac{\Phi^{-1}(0.95)\sqrt{\frac{96762}{n_F}}}{-\hat{\beta}}\right\}\right]$$

$$[100.4, \quad 624.5]$$

Note that this confidence interval uses the assumption of the parameters being normally distributed which is only true for large sample sizes. Therefore, these confidence intervals may be inaccurate. Bayesian methods must be done numerically.

Characteristics   The Birnbaum-Saunders distribution is a stochastic model of the Miner's rule.

Characteristic of α. As α decreases the distribution becomes more symmetrical around the value of β.

Hazard Rate. The hazard rate is always unimodal. The hazard rate has the following asymptotes: (Meeker & Escobar 1998, p.107)
$$h(0) = 0$$
$$\lim_{t\to\infty} h(t) = \frac{1}{2\beta\alpha^2}$$
The change point of the unimodal hazard rate for $\alpha < 0.6$ must be solved numerically, however for $\alpha > 0.6$ it can be approximated using: (Kundu et al. 2008)
$$t_c = \frac{\beta}{(-0.4604 + 1.8417\alpha)^2}$$

Lognormal and Inverse Gaussian Distribution. The shape and behavior of the Birnbaum-Saunders distribution is similar to that of the lognormal and inverse Gaussian distribution. This similarity is seen primarily in the center of the distributions. (Meeker & Escobar 1998, p.107)

Let:
$$T \sim BS(t; \alpha, \beta)$$

Scaling property (Meeker & Escobar 1998, p.107)

$$cT \sim BS(t; \alpha, c\beta)$$

where $c > 0$

Inverse property (Meeker & Escobar 1998, p.107)

$$\frac{1}{T} \sim BS\left(t; \alpha, \frac{1}{\beta}\right)$$

Applications

Fatigue-Fracture. The distribution has been designed to model crack growth to critical crack size. The model uses the Miner's rule which allows for non-constant fatigue cycles through accumulated damage. The assumption is that the crack growth during any one cycle is independent of the growth during any other cycle. The growth for each cycle has the same distribution from cycle to cycle. This is different from the proportional degradation model used to derive the lognormal distribution model, with the rate of degradation being dependent on accumulated damage. (http://www.itl.nist.gov/div898/handbook/apr/section1/apr166.htm)

Resources

Online:
http://www.itl.nist.gov/div898/handbook/eda/section3/eda366a.htm
http://www.itl.nist.gov/div898/handbook/apr/section1/apr166.htm
http://en.wikipedia.org/wiki/Birnbaum%E2%80%93Saunders_distrib
ution

Books:
Birnbaum, Z.W. & Saunders, S.C., 1969. *A New Family of Life Distributions*. Journal of Applied Probability, 6(2), 319-327.

Lemonte, A.J., Cribari-Neto, F. & Vasconcellos, K.L., 2007. *Improved statistical inference for the two-parameter Birnbaum-Saunders distribution*. Computational Statistics & Data Analysis, 51(9), 4656-4681.

Johnson, N.L., Kotz, S. & Balakrishnan, N., 1995. *Continuous Univariate Distributions, Vol. 2,* 2nd ed., Wiley-Interscience.

Rausand, M. & Høyland, A., 2004. *System reliability theory*, Wiley-IEEE.

4.3. Gamma Continuous Distribution

Probability Density Function - f(t)

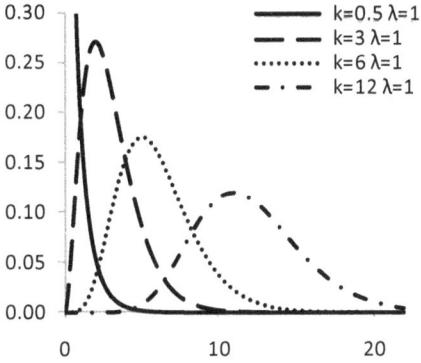

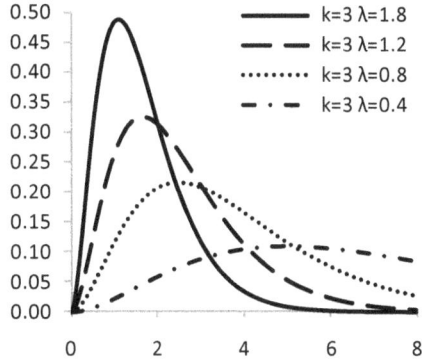

Cumulative Density Function - F(t)

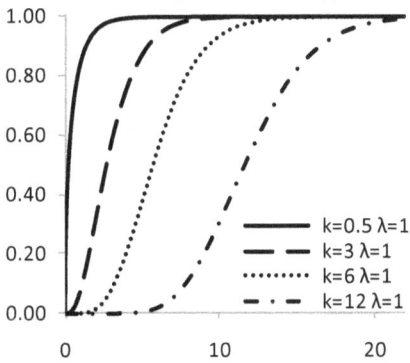

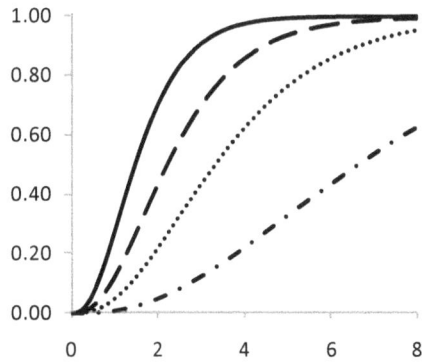

Hazard Rate - h(t)

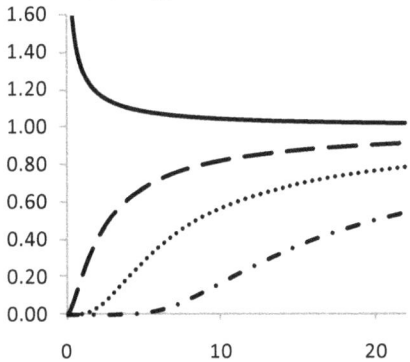

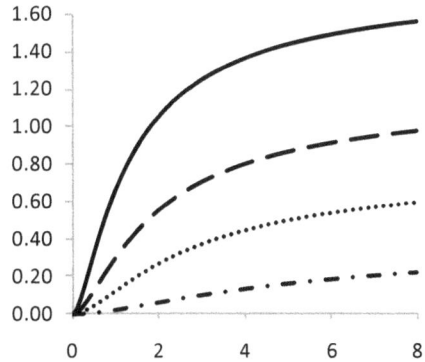

Parameters & Description

Parameters		
λ	$\lambda > 0$	*Scale Parameter:* Equal to the rate (frequency) of events/shocks. Sometimes defined as $1/\theta$ where θ is the average time between events/shocks.
k	$k > 0$	*Shape Parameter:* As an integer k can be interpreted as the number of events/shocks until failure. When not restricted to an integer, k and be interpreted as a measure of the ability to resist shocks.

Limits	$t \geq 0$

Distribution	When k is an integer (Erlang distribution)	When k is continuous

$\Gamma(k)$ is the complete gamma function. $\Gamma(k,t)$ and $\gamma(k,t)$ are the incomplete gamma functions see Section 1.6.

PDF

$$f(t) = \frac{\lambda^k t^{k-1}}{(k-1)!} e^{-\lambda t}$$

$$f(t) = \frac{\lambda^k t^{k-1}}{\Gamma(k)} e^{-\lambda t}$$

with Laplace transformation:

$$f(s) = \left(\frac{\lambda}{\lambda + s}\right)^k$$

CDF

$$F(t) = 1 - e^{-\lambda t} \sum_{n=0}^{k-1} \frac{(\lambda t)^n}{n!}$$

$$F(t) = \frac{\gamma(k, \lambda t)}{\Gamma(k)}$$
$$= \frac{1}{\Gamma(k)} \int_0^{\lambda t} x^{k-1} e^{-x}\, dx$$

Reliability

$$R(t) = e^{-\lambda t} \sum_{n=0}^{k-1} \frac{(\lambda t)^n}{n!}$$

$$R(t) = \frac{\Gamma(k, \lambda t)}{\Gamma(k)}$$
$$= \frac{1}{\Gamma(k)} \int_{\lambda t}^{\infty} x^{k-1} e^{-x}\, dx$$

Conditional Survivor Function
$P(T > x + t | T > t)$

$$e^{-\lambda x} \frac{\sum_{n=0}^{k-1} \frac{[\lambda(t+x)]^n}{n!}}{\sum_{n=0}^{k-1} \frac{(\lambda t)^n}{n!}}$$

$$m(x) = \frac{R(t+x)}{R(t)} = \frac{\Gamma(k, \lambda t + \lambda x))}{\Gamma(k, \lambda t)}$$

Where
t is the given time we know the component has survived to.
x is value of a random variable defined as the time after t. Note: $x = 0$ at t.

Mean Residual Life

$$u(t) = \frac{\int_t^{\infty} R(x) dx}{R(t)}$$

$$u(t) = \frac{\int_t^{\infty} \Gamma(k, \lambda x) dx}{\Gamma(k, \lambda t)}$$

The mean residual life does not have a closed form but has the expansion:

$$u(t) = 1 + \frac{k-1}{t} + \frac{(k-1)(k-2)}{t^2} + O(t^{-3})$$

Where $O(t^{-3})$ is Landau's notation. (Kleiber & Kotz 2003, p.161)

$$h(t) = \frac{\lambda^k t^{k-1}}{\Gamma(k) \sum_{n=0}^{k-1} \frac{(\lambda t)^n}{n!}} \qquad h(t) = \frac{\lambda^k t^{k-1}}{\Gamma(k, \lambda t)} e^{-\lambda t}$$

Hazard Rate	Series expansion of the hazard rate is: (Kleiber & Kotz 2003, p.161) $$h(t) = \left[\frac{(k-1)(k-2)}{t^2} + O(t^{-3})\right]^{-1}$$ Limits of $h(t)$ (Rausand & Høyland 2004) $$\lim_{t \to 0} h(t) = \infty \quad and \quad \lim_{t \to \infty} h(t) = \lambda \quad when\ 0 < k < 1$$ $$\lim_{t \to 0} h(t) = 0 \quad and \quad \lim_{t \to \infty} h(t) = \lambda \quad when\ k \geq 1$$
Cumulative Hazard Rate	$$H(t) = \lambda t - \ln\left[\sum_{n=0}^{k-1} \frac{(\lambda t)^n}{n!}\right] \qquad H(t) = -\ln\left[\frac{\Gamma(k, \lambda t)}{\Gamma(k)}\right]$$

Properties and Moments

Median	Numerically solve for t when: $$t_{0.5} = F^{-1}(0.5; k, \lambda)$$ or $$\gamma(k, \lambda t) = \Gamma(k, \lambda t)$$ where $\gamma(k, \lambda t)$ is the lower incomplete gamma function, see Section1.6.6.
Mode	$$\frac{k-1}{\lambda} \quad for\ k \geq 1$$ $$No\ mode\ for\ 0 < k < 1$$
Mean - 1st Raw Moment	$$\frac{k}{\lambda}$$
Variance - 2nd Central Moment	$$\frac{k}{\lambda^2}$$
Skewness - 3rd Central Moment	$2/\sqrt{k}$
Excess kurtosis - 4th Central Moment	$6/k$
Characteristic Function	$$\left(1 - \frac{it}{\lambda}\right)^{-k}$$
100α% Percentile Function	Numerically solve for t: $$t_\alpha = F^{-1}(\alpha; k, \lambda)$$

Parameter Estimation

Maximum Likelihood Function

Likelihood
Functions

$$L(k, \lambda | E) = \underbrace{\frac{\lambda^{kn_F}}{\Gamma(k)^{n_F}} \prod_{i=1}^{n_F} t_i{}^{k-1} e^{-\lambda t_i}}_{\text{complete failures}}$$

Log-Likelihood
Functions

$$\Lambda(k, \lambda | E) = kn_F \ln(\lambda) - n_F \ln\big(\Gamma(k)\big) + (k-1) \sum_{i=1}^{n_F} \ln(t_i) - \lambda \sum_{i=1}^{n_F} t_i$$

$\dfrac{\partial \Lambda}{\partial k} = 0$

$$0 = n_F \ln(\lambda) - n_F \psi(k) + \sum_{i=1}^{n_F} \{\ln(t_i)\}$$

where, $\psi(x) = \frac{d}{dx} \ln[\Gamma(x)]$ is the digamma function see Section 1.6.7.

$\dfrac{\partial \Lambda}{\partial \lambda} = 0$

$$0 = \frac{kn_F}{\lambda} - \sum_{i=1}^{n_F} t_i$$

Point
Estimates

Point estimates for $\hat{k}$ and $\hat{\lambda}$ are obtained by using numerical methods to solve the simultaneous equations above. (Kleiber & Kotz 2003, p.165)

Fisher
Information
Matrix

$$I(k, \lambda) = \begin{bmatrix} \psi'(k) & \lambda \\ \lambda & k\lambda^2 \end{bmatrix}$$

where $\psi'(x) = \frac{d^2}{dx^2} \ln\Gamma(x) = \sum_{i=0}^{\infty}(x+i)^{-2}$ is the Trigamma function. (Yang and Berger 1998, p.10)

Confidence
Intervals

For large number of samples, the Fisher information matrix can be used to estimate confidence intervals.

Bayesian

Non-informative Priors, $\pi(k, \lambda)$
(Yang and Berger 1998, p.6)

Type	Prior	Posterior
Uniform Improper Prior with limits: $\lambda \in (0, \infty)$ $k \in (0, \infty)$	1	No Closed Form
Jeffrey's Prior	$\lambda \sqrt{k.\psi'(k) - 1}$	No Closed Form
Reference Order: $\{k, \lambda\}$	$\lambda \sqrt{k.\psi'(k) - \dfrac{1}{\alpha}}$	No Closed Form
Reference Order: $\{\lambda, k\}$	$\lambda \sqrt{\psi'(k)}$	No Closed Form

where $\psi'(x) = \frac{d^2}{dx^2} ln\Gamma(x) = \sum_{i=0}^{\infty}(x+i)^{-2}$ is the Trigamma function

Conjugate Priors

UOI	Likelihood Model	Evidence	Dist. of UOI	Prior Para	Posterior Parameters
Λ from $Exp(t; \Lambda)$	Exponential	n_F failures in t_T	Gamma	k_0, λ_0	$k = k_o + n_F$ $\lambda = \lambda_o + t_T$
Λ from $Pois(k; \Lambda t)$	Poisson	n_F failures in t_T	Gamma	k_0, λ_0	$k = k_o + n_F$ $\lambda = \lambda_o + t_T$
λ where $\lambda = \alpha^{-\beta}$ from $Wbl(t; \alpha, \beta)$	Weibull with known β	n_F failures at times t_i	Gamma	k_0, λ_0	$k = k_o + n_F$ $\lambda = \lambda_o + \sum_{i=1}^{n_F} t_i^{\beta}$ (Rinne 2008, p.520)
σ^2 from $Norm(x; \mu, \sigma^2)$	Normal with known μ	n_F failures at times t_i	Gamma	k_0, λ_0	$k = k_o + n_F/2$ $\lambda = \lambda_o + \frac{1}{2} \sum_{i=1}^{n} (t_i - \mu)^2$
λ from $Gamma(x; \lambda, k)$	Gamma with known $k = k_E$	n_F failures in t_T	Gamma	η_0, Λ_0	$\eta = \eta_0 + n_F k_E$ $\Lambda = \Lambda_o + t_T$
α from $Perato(t; \theta, \alpha)$	Pareto with known θ	n_F failures at times t_i	Gamma	k_0, λ_0	$k = k_o + n_F$ $\lambda = \lambda_o + \sum_{i=1}^{n_F} \ln\left(\frac{x_i}{\theta}\right)$

where: $t_T = \sum t_i^F + \sum t_i^S = total\ time\ in\ test$

Description, Limitations and Uses

Example 1   For an example using the gamma distribution as a conjugate prior see the Poisson or Exponential distributions.

A renewal process has an exponential time between failure with parameter $\lambda = 0.01$ under the homogeneous Poisson process conditions. What is the probability the forth failure will occur before 200 hours?

$$F(200; 4, 0.01) = 0.1429$$

Example 2   5 components are put on a test with the following failure times:
38, 42, 44, 46, 55 hours.
Solving:

$$0 = \frac{5k}{\lambda} - 225$$
$$0 = 5\ln(\lambda) - 5\psi(k) + 18.9954$$

Gives:

$$\hat{k} = 21.377$$

$$\hat{\lambda} = 0.4749$$

90% confidence interval for k:

$$I(k, \lambda) = \begin{bmatrix} 0.0479 & 0.4749 \\ 0.4749 & 4.8205 \end{bmatrix}$$

$$\left[J_n(\hat{k}, \hat{\lambda})\right]^{-1} = \left[n_F I(\hat{k}, \hat{\lambda})\right]^{-1} = \begin{bmatrix} 179.979 & -17.730 \\ -17.730 & 1.7881 \end{bmatrix}$$

$$\left[\hat{k}.\exp\left\{ \frac{\Phi^{-1}(0.95)\sqrt{179.979}}{-\hat{k}} \right\}, \quad \hat{k}.\exp\left\{ \frac{\Phi^{-1}(0.95)\sqrt{179.979}}{\hat{k}} \right\} \right]$$

$$[7.6142, \quad 60.0143]$$

90% confidence interval for λ:

$$\left[\hat{\lambda}.\exp\left\{ \frac{\Phi^{-1}(0.95)\sqrt{1.7881}}{-\hat{\lambda}} \right\}, \quad \hat{\lambda}.\exp\left\{ \frac{\Phi^{-1}(0.95)\sqrt{1.7881}}{\hat{\lambda}} \right\} \right]$$

$$[0.0046, \quad 48.766]$$

Note that this confidence interval uses the assumption of the parameters being normally distributed which is only true for large sample sizes. Therefore, these confidence intervals may be inaccurate. Bayesian methods must be done numerically.

Characteristics
The gamma distribution was originally known as a Pearson Type III distribution. This distribution includes a location parameter γ which shifts the distribution along the x-axis.

$$f(t; k, \lambda, \gamma) = \frac{\lambda^k (t - \gamma)^{k-1}}{\Gamma(k)} e^{-\lambda(t-\gamma)}$$

When k is an integer, the Gamma distribution is called an Erlang distribution.

k Characteristics:
$k < 1.$ $f(0) = \infty$. There is no mode.
$k = 1.$ $f(0) = \lambda$. The gamma distribution reduces to an exponential distribution with failure rate λ. Mode at $t = 0$.
$k > 1.$ $f(0) = 0$
Large k. The gamma distribution approaches a normal distribution with $\mu = \frac{k}{\lambda}, \sigma = \sqrt{\frac{k}{\lambda^2}}$.

Homogeneous Poisson Process (HPP). Components with an exponential time to failure which undergo instantaneous renewal with an identical item undergo a HPP. The Gamma distribution is probability distribution of the kth failed item and is derived from the convolution of k exponentially distributed random variables, T_i. (See related distributions, exponential distribution).

$$T \sim Gamma(k, \lambda)$$

Scaling property:

$$aT \sim Gamma\left(k, \frac{\lambda}{a}\right)$$

Convolution property:

$$T_1 + T_2 + \dots + T_n \sim Gamma(\textstyle\sum k_i, \lambda)$$

Where λ is fixed.

Properties from (Leemis & McQueston 2008)

Renewal Theory, Homogenous Poisson Process. Used to model a renewal process where the component time to failure is exponentially distributed and the component is replaced instantaneously with a new identical component. The HPP can also be used to model ruin theory (used in risk assessments) and queuing theory.

Applications	**System Failure.** Can be used to model system failure with k backup systems.

Life Distribution. The gamma distribution is flexible in shape and can give good approximations to life data.

Bayesian Analysis. The gamma distribution is often used as a prior in Bayesian analysis to produce closed form posteriors.

Online:
http://mathworld.wolfram.com/GammaDistribution.html
http://en.wikipedia.org/wiki/Gamma_distribution
http://socr.ucla.edu/htmls/SOCR_Distributions.html (interactive web calculator)
http://www.itl.nist.gov/div898/handbook/eda/section3/eda366b.htm

Resources

Books:
Artin, E., 1964. *The Gamma Function*, New York: Holt, Rinehart & Winston.

Johnson, N.L., Kotz, S. & Balakrishnan, N., 1994. *Continuous Univariate Distributions*, Vol. 1 2nd ed., Wiley-Interscience.

Bowman, K.O. & Shenton, L.R., 1988. *Properties of estimators for the gamma distribution*, CRC Press.

Relationship to Other Distributions

Generalized Gamma Distribution

$$f(t; k, \lambda, \gamma, \xi) = \frac{\xi \lambda^{\xi k} (t - \gamma)^{\xi k - 1}}{\Gamma(k)} \exp\{-[\lambda(t - \gamma)]^k\}$$

$Gamma(t; k, \lambda, \gamma, \xi)$

λ - Scale Parameter
k - Shape Parameter
γ - Location parameter
ξ - Second shape parameter

The generalized gamma distribution has been derived because it is a generalization of a large amount of probability distributions. Such as:

$$Gamma(t; 1, \lambda, 0, 1) = Exp(t; \lambda)$$

$$Gamma\left(t; 1, \frac{1}{\mu}, \beta, 1\right) = Exp(t; \mu, \beta)$$

$$Gamma\left(t; 1, \frac{1}{\alpha}, 0, \beta\right) = Weibull(t; \alpha, \beta)$$

$$Gamma\left(t; 1, \frac{1}{\alpha}, \gamma, \beta\right) = Weibull(t; \alpha, \beta, \gamma)$$

$$Gamma\left(t; \frac{n}{2}, \frac{1}{2}, 0, 1\right) = \chi^2(t; n)$$

$$Gamma\left(t; \frac{n}{2}, \frac{1}{\sqrt{2}}, 0, 2\right) = \chi(t; n)$$

$$Gamma\left(t; 1, \frac{1}{\sigma}, 0, 2\right) = Rayleigh(t; \sigma)$$

Exponential Distribution

$Exp(t; \lambda)$

Let, $T_1 \ldots T_k \sim Exp(\lambda)$ and $T_t = T_1 + T_2 + \cdots + T_k$
Then,

$$T_t \sim Gamma(k, \lambda)$$

This gives the Gamma distribution its convolution property.

Special Case:

$$Exp(t; \lambda) = Gamma(t; k = 1, \lambda)$$

Let, $T_1 \ldots T_k \sim Exp(\lambda)$ and $T_t = T_1 + T_2 + \cdots + T_k$
Then,

$$T_t \sim Gamma(k, \lambda)$$

Poisson Distribution

$Pois(k; \lambda t)$

The Poisson distribution is the probability that exactly k failures have been observed in time t. This is the probability that t is between T_k and T_{k+1}.

$$f_{Poisson}(k; \lambda t) = \int_k^{k+1} f_{Gamma}(t; x, \lambda) dx$$
$$= F_{Gamma}(t; k + 1, \lambda) - F_{Gamma}(t; k, \lambda)$$

where k is an integer.

Normal Distribution

$Norm(t; \mu, \sigma)$

Special Case for large k:

$$\lim_{k \to \infty} Gamma(k, \lambda) = Norm\left(\mu = \frac{k}{\lambda}, \sigma = \sqrt{\frac{k}{\lambda^2}}\right)$$

Chi-square Distribution

$\chi^2(t; v)$

Special Case:

$$\chi^2(t; v) = Gamma(t; k = \frac{v}{2}, \lambda = \frac{1}{2})$$

where v is an integer

Inverse Gamma Distribution $IG(t; \alpha, \beta)$	Let, $X \sim Gamma(k, \lambda)$ $\qquad$ and $\qquad Y = \frac{1}{X}$ Then, $$Y \sim IG(\alpha = k, \beta = \lambda)$$
Beta Distribution $Beta(t; \alpha, \beta)$	Let, $X_1, X_2 \sim Gamma(k_i, \lambda_i)$ $\qquad$ and $\qquad Y = \frac{X_1}{X_1 + X_2}$ Then, $$Y \sim Beta(\alpha = k_1, \beta = k_2)$$
Dirichlet Distribution $Dir_d(x; \alpha)$	Let, $Y_i \sim Gamma(\lambda, k_i)$ $\;i.i.d$ and $\;V = \sum_{i=1}^{d} Y_i$ Then, $$V \sim Gamma(\lambda, \sum k_i)$$ Let: $$Z = \left[\frac{Y_1}{V}, \frac{Y_2}{V}, \dots, \frac{Y_d}{V} \right]$$ Then: $$Z \sim Dir_d(\alpha_1, \dots, \alpha_k)$$ *i.i.d: independent and identically distributed
Wishart Distribution $Wishart_d(n; \Sigma)$	The Wishart Distribution is the multivariate generalization of the gamma distribution.

4.4. Logistic Continuous Distribution

Probability Density Function - f(t)

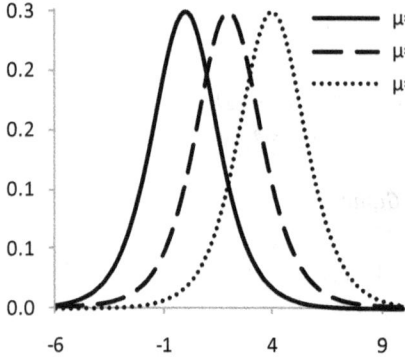

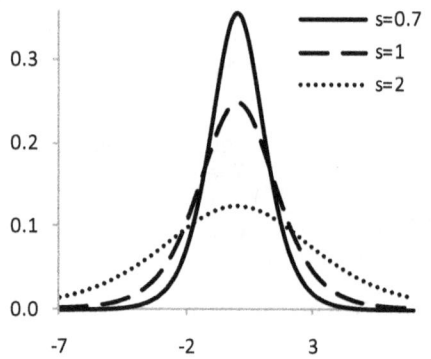

Cumulative Density Function - F(t)

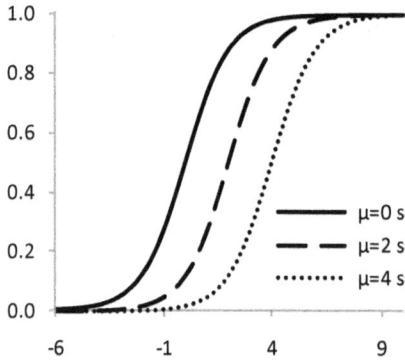

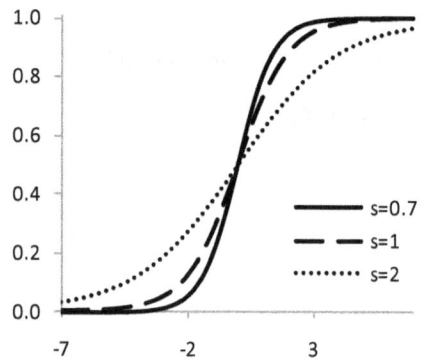

Hazard Rate - h(t)

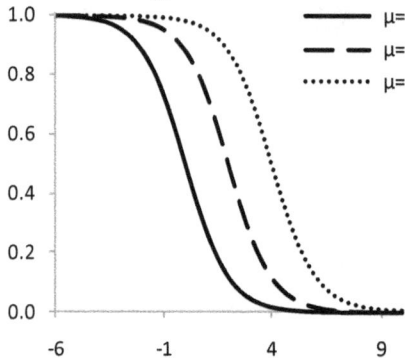

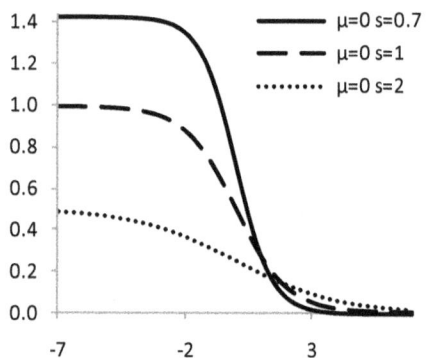

Parameters & Description

Parameters			
	μ	$-\infty < \mu < \infty$	Location parameter. μ is the mean, median and mode of the distribution.
	s	$s > 0$	Scale parameter. Proportional to the standard deviation of the distribution.
Limits		$-\infty < t < \infty$	

Distribution	Formulas
PDF	$$f(t) = \frac{e^z}{s(1 + e^z)^2} = \frac{e^{-z}}{s(1 + e^{-z})^2}$$ $$= \frac{1}{4s} \operatorname{sech}^2\left(\frac{t - \mu}{2s}\right)$$ where $$z = \frac{t - \mu}{s}$$
CDF	$$F(t) = \frac{1}{1 + e^{-z}} = \frac{e^z}{1 + e^z}$$ $$= \frac{1}{2} + \frac{1}{2}\tanh\left(\frac{t - \mu}{2s}\right)$$
Reliability	$$R(t) = \frac{1}{1 + e^z}$$
Conditional Survivor Function $P(T > x + t \| T > t)$	$$m(x) = R(x\|t) = \frac{R(t + x)}{R(t)} = \frac{1 + \exp\left\{\frac{t - \mu}{s}\right\}}{1 + \exp\left\{\frac{t + x - \mu}{s}\right\}}$$ where, t is the given time we know the component has survived to. x is value of a random variable defined as the time after t. Note: $x = 0$ at t.
Mean Residual Life	$$u(t) = (1 + e^z)\left(s.\ln\left[e^{t/s} + e^{\mu/s}\right] - t\right)$$
Hazard Rate	$$h(t) = \frac{1}{s(1 + e^{-z})} = \frac{F(t)}{s}$$ $$= \frac{1}{s + s\exp\left\{\frac{\mu - t}{s}\right\}}$$
Cumulative Hazard Rate	$$H(t) = \ln\left[1 + \exp\left\{\frac{t - \mu}{s}\right\}\right]$$

Properties and Moments

Median	μ

Mode	μ
Mean - 1st Raw Moment	μ
Variance - 2nd Central Moment	$\dfrac{\pi^2}{3}s^2$
Skewness - 3rd Central Moment	0
Excess kurtosis - 4th Central Moment	$\dfrac{6}{5}$
Characteristic Function	$e^{i\mu t}B(1-ist,1+ist)$ for $|st|<1$
100γ % Percentile Function	$t_\gamma = \mu + s\ln\left(\dfrac{\gamma}{1-\gamma}\right)$

Parameter Estimation

Plotting Method

Least Mean Square	X-Axis	Y-Axis	
$y = mx + c$	t_i	$\ln[F] - \ln[1-F]$	$\hat{s} = \dfrac{1}{m}$
			$\hat{\mu} = -c\hat{s}$

Maximum Likelihood Function

Likelihood Function

For complete data:

$$L(\mu,s|E) = \prod_{i=1}^{n_F} \underbrace{\frac{\exp\left\{\frac{t_i-\mu}{-s}\right\}}{s\left(1+\exp\left\{\frac{t_i-\mu}{-s}\right\}\right)^2}}_{\text{complete failures}}$$

Log-Likelihood Function

$$\Lambda(\mu,s|E) = -n_F\ln s + \underbrace{\sum_{i=1}^{n_F}\left\{\frac{t_i-\mu}{-s}\right\} - 2\sum_{i=1}^{n_F}\ln\left(1+\exp\left\{\frac{t_i-\mu}{-s}\right\}\right)}_{\text{complete failures}}$$

$\dfrac{\partial\Lambda}{\partial\mu} = 0$

$$\frac{\partial\Lambda}{\partial\mu} = \frac{n_F}{s} - \underbrace{\frac{2}{s}\sum_{i=1}^{n_F}\frac{1}{\left(1+\exp\left\{\frac{t_i-\mu}{s}\right\}\right)}}_{\text{complete failures}} = 0$$

$\dfrac{\partial\Lambda}{\partial s} = 0$

$$\frac{\partial\Lambda}{\partial s} = -\frac{n_F}{s} - \underbrace{\frac{1}{s}\sum_{i=1}^{n_F}\left(\frac{t_i-\mu}{s}\right)\left[\frac{1-\exp\left\{\frac{t_i-\mu}{s}\right\}}{1+\exp\left\{\frac{t_i-\mu}{s}\right\}}\right]}_{\text{complete failures}} = 0$$

MLE Point Estimates

The MLE estimates for $\hat{\mu}$ and $\hat{s}$ are found by solving the following equations:

$$\frac{1}{2} - \frac{1}{n_F}\sum_{i=1}^{n_F}\left[1+\exp\left\{\frac{t_i-\mu}{s}\right\}\right]^{-1} = 0$$

$$1 + \frac{1}{n_F} \sum_{i=1}^{n_F} \left(\frac{t_i - \mu}{s}\right) \frac{1 - \exp\left\{\frac{t_i - \mu}{s}\right\}}{1 + \exp\left\{\frac{t_i - \mu}{s}\right\}} = 0$$

These estimates are biased. (Balakrishnan 1991) provides tables derived from Monte Carlo simulation to correct the bias.

Fisher
Information

$$I(\mu, s) = \begin{bmatrix} \frac{1}{3s^2} & 0 \\ 0 & \frac{\pi^2 + 3}{9s^2} \end{bmatrix}$$

(Antle et al. 1970)

$100\gamma\%$
Confidence
Intervals

Confidence intervals are most often obtained from tables derived from Monte Carlo simulation. Corrections from using the Fisher Information matrix method are given in (Antle et al. 1970).

Bayesian
Non-informative Priors $\pi_0(\mu, s)$

Type	Prior
Jeffery Prior	$\frac{1}{s}$

Description, Limitations and Uses

Example

The accuracy of a cutting machine used in manufacturing is desired to be measured. 5 cuts at the required length are made and measured as:
$$7.436, 10.270, 10.466, 11.039, 11.854 \ mm$$

Numerically solving MLE equations gives:
$$\hat{\mu} = 10.446$$
$$\hat{s} = 0.815$$

This gives a mean of 10.446 and a variance of 2.183. Compared to the same data used in the Normal distribution section it can be seen that this estimate is very similar to a normal distribution.

90% confidence interval for μ:
$$\left[\hat{\mu} - \Phi^{-1}(0.95)\sqrt{\frac{3\hat{s}^2}{n_F}}, \quad \hat{\mu} + \Phi^{-1}(0.95)\sqrt{\frac{3\hat{s}^2}{n_F}}\right]$$
$$[9.408, \quad 11.4844]$$

90% confidence interval for s:

$$\left[\hat{s}.\exp\left\{ \frac{\Phi^{-1}(0.95)\sqrt{\frac{9\hat{s}^2}{n_F(3+\pi^2)}}}{-\hat{s}} \right\}, \quad \hat{s}.\exp\left\{ \frac{\Phi^{-1}(0.95)\sqrt{\frac{9\hat{s}^2}{n_F(3+\pi^2)}}}{\hat{s}} \right\} \right]$$

$$[0.441, \quad 1.501]$$

Note that this confidence interval uses the assumption of the parameters being normally distributed which is only true for large sample sizes. Therefore, these confidence intervals may be inaccurate.

Bayesian methods must be calculated using numerical methods.

Characteristics
The logistic distribution is most often used to model growth rates (and has been used extensively in biology and chemical applications). In reliability engineering it is most often used as a life distribution. Its CDF also used as the model for probability of detection of damage or flaws in non-destructive testing.

Shape. There is no shape parameter and so the logistic distribution is always a bell-shaped curve. Increasing μ shifts the curve to the right, increasing s increases the spread of the curve.

Normal Distribution. The shape of the logistic distribution is very similar to that of a normal distribution with the logistic distribution having slightly 'longer tails'. It would take a large number of samples to distinguish between the distributions. The main difference is that the hazard rate approaches $1/s$ for large t. The logistic function has historically been preferred over the normal distribution because of its simplified form. (Meeker & Escobar 1998, p.89)

Alternative Parameterization. It is equally as popular to present the logistic distribution using the true standard deviation $\sigma = \pi s/\sqrt{3}$. This form is used in reference book, Balakrishnan 1991, and gives the following cdf:

$$F(t) = \frac{1}{1 + \exp\left\{ \frac{-\pi}{\sqrt{3}}\left(\frac{t-\mu}{\sigma} \right) \right\}}$$

Standard Logistic Distribution. The standard logistic distribution has $\mu = 0, s = 1$. The standard logistic distribution random variable, Z, is related to the logistic distribution:

$$Z = \frac{X-\mu}{s}$$

Let:

$$T \sim Logistic(t; \mu, s)$$

Scaling property (Leemis & McQueston 2008)
$$aT \sim Logistic(t; \mu, as)$$

Rate Relationships. The distribution has the following rate relationships which make it suitable for modeling growth (Hastings et al. 2000, p.127):

$$h(t) = \frac{f(t)}{R(t)} = \frac{F(t)}{s}$$

$$z = \ln\left[\frac{F(t)}{R(t)}\right] = \ln[F(t)] - \ln[1 - F(t)]$$

where

$$z = \frac{t - \mu}{s}$$

when $\mu = 0$ and $s = 1$:

$$f(t) = \frac{dF(t)}{dt} = F(t)R(t)$$

Applications

Growth Model. The logistic distribution most common use is a growth model.

Probability of Detection. The cdf of logistic distribution is commonly used to represent the probability of detection damaged materials sensors and detection instruments. For example, probability of detection of embedded flaws in metals using ultrasonic signals.

Life Distribution. In reliability applications it is used as a life distribution. It is similar in shape to a normal distribution and so is often used instead of a normal distribution due to its simplified form. (Meeker & Escobar 1998, p.89)

Logistic Regression. Logistic regression is a generalized linear regression model used to predict binary outcomes. (Agresti 2002)

Resources

Online:
http://mathworld.wolfram.com/LogisticDistribution.html
http://en.wikipedia.org/wiki/Logistic_distribution
http://socr.ucla.edu/htmls/SOCR_Distributions.html (web calc)
http://www.weibull.com/LifeDataWeb/the_logistic_distribution.htm

Books:
Balakrishnan, 1991. Handbook of the Logistic Distribution 1st ed., CRC.
Johnson, N.L., Kotz, S. & Balakrishnan, N., 1995. Continuous Univariate Distributions, Vol. 2 2nd ed., Wiley-Interscience.

Relationship to Other Distributions

Let

Exponential
Distribution
$Exp(t; \lambda)$

$$X \sim Exp(\lambda = 1) \qquad and \qquad Y = \ln\left\{\frac{e^{-X}}{1 + e^{-X}}\right\}$$

Then

$$Y \sim Logistic(0,1)$$

(Hastings et al. 2000, p.127)

Let

Pareto
Distribution
$Pareto(\theta, \alpha)$

$$X \sim Pareto(\theta, \alpha) \qquad and \qquad Y = -\ln\left\{\left(\frac{X}{\theta}\right)^\alpha - 1\right\}$$

Then

$$Y \sim Logistic(0,1)$$

(Hastings et al. 2000, p.127)

Let

Gumbel
Distribution
$Gumbel(\alpha, \beta)$

$$X_i \sim Gumbel(\alpha, \beta) \qquad and \qquad Y = X_1 - X_2$$

Then

$$Y \sim Logistic(0, \beta)$$

(Hastings et al. 2000, p.127)

4.5. Normal (Gaussian) Continuous Distribution

Probability Density Function - f(t)

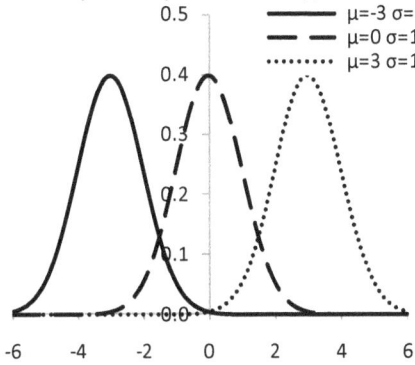

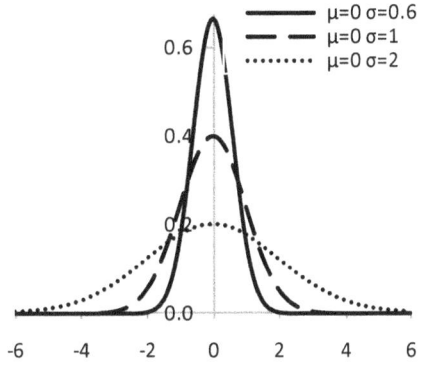

Cumulative Density Function - F(t)

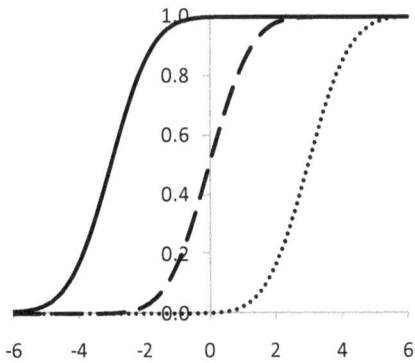

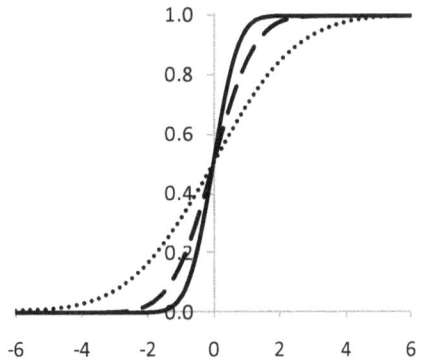

Hazard Rate - h(t)

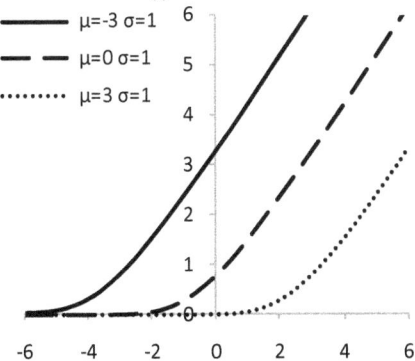

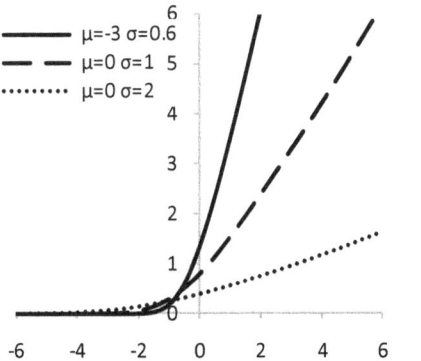

Parameters & Description

Parameters			
	μ	$-\infty < \mu < \infty$	*Location parameter:* The mean of the distribution.
	σ^2	$\sigma^2 > 0$	*Scale parameter:* The standard deviation of the distribution.

Limits	$-\infty < t < \infty$

Distribution — Formulas

PDF

$$f(t) = \frac{1}{\sigma\sqrt{2\pi}} \exp\left[-\frac{1}{2}\left(\frac{t-\mu}{\sigma}\right)^2\right]$$

$$= \frac{1}{\sigma}\phi\left[\frac{t-\mu}{\sigma}\right]$$

where ϕ is the standard normal pdf with $\mu = 0$ and $\sigma^2 = 1$.

CDF

$$F(t) = \frac{1}{\sigma\sqrt{2\pi}} \int_{-\infty}^{t} \exp\left[-\frac{1}{2}\left(\frac{\theta-\mu}{\sigma}\right)^2\right] d\theta$$

$$= \frac{1}{2} + \frac{1}{2}\text{erf}\left(\frac{t-\mu}{\sigma\sqrt{2}}\right)$$

$$= \Phi\left(\frac{t-\mu}{\sigma}\right)$$

where, Φ is the standard normal cdf with $\mu = 0$ and $\sigma^2 = 1$.

Reliability

$$R(t) = 1 - \Phi\left(\frac{t-\mu}{\sigma}\right)$$

$$= \Phi\left(\frac{\mu-t}{\sigma}\right)$$

Conditional Survivor Function $P(T > x + t | T > t)$

$$m(x) = R(x|t) = \frac{R(t+x)}{R(t)} = \frac{\Phi\left(\frac{\mu-x-t}{\sigma}\right)}{\Phi\left(\frac{\mu-t}{\sigma}\right)}$$

Where
t is the given time we know the component has survived to.
x is a random variable defined as the time after t. Note: $x = 0$ at t.

Mean Residual Life

$$u(t) = \frac{\int_t^\infty R(x)dx}{R(t)} = \frac{\int_t^\infty R(x)dx}{R(t)}$$

Hazard Rate

$$h(t) = \frac{\phi\left[\frac{t-\mu}{\sigma}\right]}{\sigma\left(\Phi\left[\frac{\mu-t}{\sigma}\right]\right)}$$

Cumulative Hazard Rate

$$H(t) = -\ln\left[\Phi\left(\frac{\mu-t}{\sigma}\right)\right]$$

Properties and Moments

Median	μ

Mode	μ
Mean - 1st Raw Moment	μ
Variance - 2nd Central Moment	σ^2
Skewness - 3rd Central Moment	0
Excess kurtosis - 4th Central Moment	0
Characteristic Function	$\exp\left(i\mu t - \frac{1}{2}\sigma^2 t^2\right)$
100α% Percentile Function	$t_\alpha = \mu + \sigma\Phi^{-1}(\alpha)$ $= \mu + \sigma\sqrt{2}\,\mathrm{erf}^{-1}(2\alpha - 1)$

Parameter Estimation

Plotting Method

	X-Axis	Y-Axis	
Least Mean Square $y = mx + c$	t_i	$invNorm[F(t_i)]$	$\hat{\mu} = -\dfrac{c}{m}$ $\hat{\sigma} = \dfrac{1}{m}, \quad \widehat{\sigma^2} = \dfrac{1}{m^2}$

Maximum Likelihood Function

Likelihood Function

For complete data:

$$L(\mu, \sigma|E) = \underbrace{\frac{1}{(\sigma\sqrt{2\pi})^{n_F}}\prod_{i=1}^{n_F}\exp\left(-\frac{1}{2}\left[\frac{t_i - \mu}{\sigma}\right]^2\right)}_{\text{complete failures}}$$

$$= \underbrace{\frac{1}{(\sigma\sqrt{2\pi})^{n_F}}\exp\left(-\frac{1}{2\sigma^2}\sum_{i=1}^{n_F}(t_i - \mu)^2\right)}_{\text{complete failures}}$$

Log-Likelihood Function

$$\Lambda(\mu, \sigma|E) = \underbrace{-n_F\ln\left(\sigma\sqrt{2\pi}\right) - \frac{1}{2\sigma^2}\sum_{i=1}^{n_F}(t_i - \mu)^2}_{\text{complete failures}}$$

$\dfrac{\partial\Lambda}{\partial\mu} = 0$

solve for μ to get MLE $\hat{\mu}$:

$$\frac{\partial\Lambda}{\partial\mu} = \underbrace{\frac{\mu n_F}{\sigma^2} - \frac{1}{\sigma^2}\sum_{i=1}^{n_F}t_i = 0}_{\text{complete failures}}$$

$\dfrac{\partial\Lambda}{\partial\sigma} = 0$

solve for σ to get $\hat{\sigma}$:

$$\frac{\partial\Lambda}{\partial\sigma} = \underbrace{-\frac{n_F}{\sigma} + \frac{1}{\sigma^3}\sum_{i=1}^{n_F}(t_i - \mu)^2 = 0}_{\text{complete failures}}$$

MLE Point Estimates

When there is only complete failure data the point estimates can be given as:

$$\hat{\mu} = \frac{1}{n_F}\sum_{i=1}^{n_F} t_i \qquad \widehat{\sigma^2} = \frac{1}{n_F}\sum_{i=1}^{n_F} (t_i - \mu)^2$$

In most cases the unbiased estimators are used:

$$\hat{\mu} = \frac{1}{n_F}\sum_{i=1}^{n_F} t_i \qquad \widehat{\sigma^2} = \frac{1}{n_F - 1}\sum_{i=1}^{n_F} (t_i - \mu)^2$$

Fisher Information		$I(\mu, \sigma^2) = \begin{bmatrix} 1/\sigma^2 & 0 \\ 0 & -1/2\sigma^4 \end{bmatrix}$		
100γ% Confidence Intervals		1 Sided - Lower	2 Sided - Lower	2 Sided - Upper
(for complete data)	μ	$\hat{\mu} - \frac{\hat{\sigma}}{\sqrt{n}} t_\gamma(n-1)$	$\hat{\mu} - \frac{\hat{\sigma}}{\sqrt{n}} t_{\left(\frac{1+\gamma}{2}\right)}(n-1)$	$\hat{\mu} + \frac{\hat{\sigma}}{\sqrt{n}} t_{\left(\frac{1+\gamma}{2}\right)}(n-1)$
	σ^2	$\widehat{\sigma^2}\dfrac{(n-1)}{\chi_\alpha^2(n-1)}$	$\widehat{\sigma^2}\dfrac{(n-1)}{\chi_{\left(\frac{1+\gamma}{2}\right)}^2(n-1)}$	$\widehat{\sigma^2}\dfrac{(n-1)}{\chi_{\left(\frac{1-\gamma}{2}\right)}^2(n-1)}$

(Nelson 1982, pp.218-220) Where $t_\gamma(n-1)$ is the $100\gamma^{th}$ percentile of the t-distribution with $n-1$ degrees of freedom and $\chi_\gamma^2(n-1)$ is the $100\gamma^{th}$ percentile of the χ^2-distribution with $n-1$ degrees of freedom.

Bayesian

Non-informative Priors when σ^2 is known, $\pi_0(\mu)$
(Yang and Berger 1998, p.22)

Type	Prior	Posterior
Uniform Proper Priors with limits $\mu \in [a,b]$	$\dfrac{1}{b-a}$	Truncated Normal Distribution For $a \le \mu \le b$ $c.Norm\left(\mu; \dfrac{\sum_{i=1}^{n_F} t_i^F}{n_F}, \dfrac{\sigma^2}{n_F}\right)$ Otherwise, $\pi(\mu) = 0$
All	1	$Norm\left(\mu; \dfrac{\sum_{i=1}^{n_F} t_i^F}{n_F}, \dfrac{\sigma^2}{n_F}\right)$ when $\mu \in (\infty, \infty)$

Non-informative Priors when μ is known, $\pi_0(\sigma^2)$
(Yang and Berger 1998, p.23)

Type	Prior	Posterior
Uniform Proper Prior with limits $\sigma^2 \in [a,b]$	$\dfrac{1}{b-a}$	Truncated Inverse Gamma Distribution For $a \le \sigma^2 \le b$ $c.IG\left(\sigma^2; \dfrac{(n_F - 2)}{2}, \dfrac{S^2}{2}\right)$ Otherwise $\pi(\sigma^2) = 0$

Uniform Improper Prior with limits $\sigma^2 \in (0, \infty)$	1	$IG\left(\sigma^2; \dfrac{(n_F - 2)}{2}, \dfrac{S^2}{2}\right)$ See section 1.7.1
Jeffery's, Reference, MDIP Prior	$\dfrac{1}{\sigma^2}$	$IG\left(\sigma^2; \dfrac{n_F}{2}, \dfrac{S^2}{2}\right)$ with limits $\sigma^2 \in (0, \infty)$ See section 1.7.1

Non-informative Priors when μ and σ^2 are unknown, $\pi_o(\mu, \sigma^2)$
(Yang and Berger 1998, p.23)

Type	Prior	Posterior
Improper Uniform with limits: $\mu \in (\infty, \infty)$ $\sigma^2 \in (0, \infty)$	1	$\pi(\mu|E) \sim T\left(\mu; n_F - 3, \bar{t}, \dfrac{S^2}{n_F(n_F - 3)}\right)$ See Section 1.7.2 $\pi(\sigma^2|E) \sim IG\left(\sigma^2; \dfrac{(n_F - 3)}{2}, \dfrac{S^2}{2}\right)$ See Section 1.7.1
Jeffery's Prior	$\dfrac{1}{\sigma^4}$	$\pi(\mu|E) \sim T\left(\mu; n_F + 1, \bar{t}, \dfrac{S^2}{n_F(n_F + 1)}\right)$ when $\mu \in (\infty, \infty)$ See Section 1.7.2 $\pi(\sigma^2|E) \sim IG\left(\sigma^2; \dfrac{(n_F + 1)}{2}, \dfrac{S^2}{2}\right)$ when $\sigma^2 \in (0, \infty)$ See Section 1.7.1
Reference Prior ordering $\{\phi, \sigma\}$	$\pi_o(\phi, \sigma^2)$ $\propto \dfrac{1}{\sigma\sqrt{2 + \phi^2}}$ where $\phi = \mu/\sigma$	No Closed Form
Reference where μ and σ^2 are separate groups. MDIP Prior	$\dfrac{1}{\sigma^2}$	$\pi(\mu|E) \sim T\left(\mu; n_F - 1, \bar{t}, \dfrac{S^2}{n_F(n_F - 1)}\right)$ when $\mu \in (\infty, \infty)$ See Section 1.7.2 $\pi(\sigma^2|E) \sim IG\left(\sigma^2; \dfrac{(n_F - 1)}{2}, \dfrac{S^2}{2}\right)$ when $\sigma^2 \in (0, \infty)$ See Section 1.7.1

where

$$S^2 = \sum_{i=1}^{n_F} (t_i - \bar{t})^2 \quad \text{and} \quad \bar{t} = \frac{1}{n_F} \sum_{i=1}^{n_F} t_i$$

Conjugate Priors

UOI	Likelihood Model	Evidence	Dist. of UOI	Prior Para	Posterior Parameters
μ from $Norm(t; \mu, \sigma^2)$	Normal with known σ^2	n_F failures at times t_i	Normal	u_o, v_0	$u = \dfrac{\dfrac{u_0}{v_0} + \dfrac{\sum_{i=1}^{n_F} t_i^F}{\sigma^2}}{\dfrac{1}{v_0} + \dfrac{n_F}{\sigma^2}}$ $v = \dfrac{1}{\dfrac{1}{v_0} + \dfrac{n_F}{\sigma^2}}$
σ^2 from $Norm(t; \mu, \sigma^2)$	Normal with known μ	n_F failures at times t_i	Gamma	k_0, λ_0	$k = k_o + n_F/2$ $\lambda = \lambda_o + \dfrac{1}{2} \sum_{i=1}^{n_F} (t_i - \mu)^2$
μ_N from $LogN(t; \mu_N, \sigma_N^2)$	Lognormal with known σ_N^2	n_F failures at times t_i	Normal	u_o, v_0	$u = \dfrac{\dfrac{u_0}{\sigma_0^2} + \dfrac{\sum_{i=1}^{n_F} \ln(t_i)}{\sigma_N^2}}{\dfrac{1}{v^2} + \dfrac{n_F}{\sigma_N^2}}$ $v = \dfrac{1}{\dfrac{1}{v^2} + \dfrac{n_F}{\sigma_N^2}}$

Description, Limitations and Uses

Example

The accuracy of a cutting machine used in manufacturing is desired to be measured. 5 cuts at the required length are made and measured as:

$$7.436, 10.270, 10.466, 11.039, 11.854 \; mm$$

MLE Estimates are:

$$\hat{\mu} = \frac{\sum t_i^F}{n_F} = 10.213$$

$$\widehat{\sigma^2} = \frac{\sum (t_i^F - \hat{\mu}_t)^2}{n_F - 1} = 2.789$$

90% confidence interval for μ:

$$\left[\hat{\mu} - \frac{\hat{\sigma}}{\sqrt{5}} t_{\{0.95\}}(4), \quad \hat{\mu} + \frac{\hat{\sigma}}{\sqrt{5}} t_{\{0.95\}}(4) \right]$$
$$[10.163, \quad 10.262]$$

90% confidence interval for σ^2:

$$\left[\widehat{\sigma^2} \frac{4}{\chi_{\{0.95\}}^2(4)}, \quad \widehat{\sigma^2} \frac{4}{\chi_{\{0.05\}}^2(4)} \right]$$
$$[1.176, \quad 15.697]$$

A Bayesian point estimate using the Jeffery non-informative improper prior $1/\sigma^4$ with posterior for $\mu \sim T(6, 10.213, 0.558)$ and $\sigma^2 \sim IG(3, 5.578)$ has point estimates:

$$\hat{\mu} = E[T(6,6.595,0.412)] = \mu = 10.213$$

$$\widehat{\sigma^2} = E[IG(3,5.578)] = \frac{5.578}{2} = 2.789$$

With 90% confidence intervals:

μ

$$[F_T^{-1}(0.05) = 8.761, \quad F_T^{-1}(0.95) = 11.665]$$

σ^2

$$[1/F_G^{-1}(0.95) = 0.886, \quad 1/F_G^{-1}(0.05) = 6.822]$$

Characteristics   Also known as a Gaussian distribution or bell curve.

Unit Normal Distribution. Also known as the standard normal distribution is when $\mu = 0$ and $\sigma = 1$ with pdf $\phi(z)$ and cdf $\Phi(z)$. If X is normally distributed with mean μ and standard deviation σ then the following transformation is used:

$$z = \frac{x - \mu}{\sigma}$$

Central Limit Theorem. Let $X_1, X_2, \ldots, X_n$ be a sequence of n independent and identically distributed (i.i.d) random variables each having a mean of μ and a variance of σ^2. As the sample size increases, the distribution of the sample average of these random variables approaches the normal distribution with mean μ and variance σ^2/n irrespective of the shape of the original distribution. Formally:

$$S_n = X_1 + \cdots + X_n$$

If we define new random variables:

$$Z_n = \frac{S_n - n\mu}{\sigma\sqrt{n}}, \quad and \quad Y = \frac{S_n}{n}$$

The distribution of Z_n converges to the standard normal distribution. The distribution of S_n converges to a normal distribution with mean μ and standard deviation of $\sigma/\sqrt{n}$.

Sigma Intervals. Often intervals of the normal distribution are expressed in terms of distance away from the mean in units of sigma. The following is approximate values for each sigma:

Interval	$\Phi(\mu + n\sigma) - \Phi(\mu - n\sigma)$
$\mu \pm \sigma$	68.2689492137%
$\mu \pm 2\sigma$	95.4499736104%
$\mu \pm 3\sigma$	99.7300203937%
$\mu \pm 4\sigma$	99.9936657516%
$\mu \pm 5\sigma$	99.9999426697%
$\mu \pm 6\sigma$	99.9999998027%

Truncated Normal. Often in reliability engineering a truncated normal distribution may be used due to the limitation that $t \geq 0$. See Truncated Normal Continuous Distribution.

Inflection Points:
Inflection points occur one standard deviation away from the mean $(\mu \pm \sigma)$.

Mean / Median / Mode:
The mean, median and mode are always equal to μ.

Hazard Rate. The hazard rate is increasing for all t. The Standard Normal Distribution's hazard rate approaches $h(t) = t$ as t becomes large.

Let:

$$X \sim Norm(\mu, \sigma^2)$$

Convolution Property

$$\sum_{i=1}^{n} X_i \sim Norm\left(\sum \mu_i, \sum \sigma_i^2\right)$$

Scaling Property

$$aX + b \sim Norm(a\mu + b, a^2\sigma^2)$$

Linear Combination Property:

$$\sum_{i=1}^{n} a_i X_i + b_i \sim Norm\left(\sum \{a_i \mu_i + b_i\}, \sum \{a_i^2 \sigma_i^2\}\right)$$

Applications

Approximations to Other Distributions. The origin of the Normal Distribution was from an approximation of the Binomial distribution. Due to the Central Limit Theory, the Normal distribution can be used to approximate many distributions as detailed under 'Related Distributions'.

Strength-Stress Interference. When the strength of a component follows a distribution and the stress that component is subjected to follows a distribution there exists a probability that the stress will be greater than the strength. When both distributions are a normal distribution, there is a closed form solution to the interference probability.

Life Distribution. When used as a life distribution a truncated Normal Distribution may be used due to the constraint $t \geq 0$. However, it is often found that the difference in results is negligible. (Rausand & Høyland 2004)

Time Distributions. The normal distribution may be used to model simple repair or inspection tasks that have a typical duration with variation which is symmetrical about the mean. This is typical for inspection and preventative maintenance times.

Analysis of Variance (ANOVA). A test used to analyze variance and dependence of variables. A popular model used to conduct ANOVA assumes the data comes from a normal population.

Six Sigma Quality Management. Six Sigma is a business management strategy which aims to reduce costs in manufacturing processes by removing variance in quality (defects). Current manufacturing standards aim for an expected 3.4 defects out of one million parts: $2\Phi(-6)$. (Six Sigma Academy 2009)

Resources

Online:
http://www.weibull.com/LifeDataWeb/the_normal_distribution.htm
http://mathworld.wolfram.com/NormalDistribution.html
http://en.wikipedia.org/wiki/Normal_distribution
http://socr.ucla.edu/htmls/SOCR_Distributions.html (web calc)

Books:
Patel, J.K. & Read, C.B., 1996. *Handbook of the Normal Distribution* 2nd ed., CRC.

Simon, M.K., 2006. *Probability Distributions Involving Gaussian Random Variables: A Handbook for Engineers and Scientists*, Springer.

Relationship to Other Distributions

Truncated Normal Distribution

$TNorm(x; \mu, \sigma, a_L, b_U)$

Let:
$$X \sim Norm(\mu, \sigma^2)$$
$$X \in (\infty, \infty)$$
Then:
$$Y \sim TNorm(\mu, \sigma^2, a_L, b_U)$$
$$Y \in [a_L, b_U]$$

Lognormal Distribution

$LogN(t; \mu_N, \sigma_N^2)$

Let:
$$X \sim LogN(\mu_N, \sigma_N^2)$$
$$Y = \ln(X)$$
Then:
$$Y \sim Norm(\mu, \sigma^2)$$
Where:
$$\mu_N = \ln\left(\frac{\mu^2}{\sqrt{\sigma^2 + \mu^2}}\right), \quad \sigma_N = \sqrt{\ln\left(\frac{\sigma^2 + \mu^2}{\mu^2}\right)}$$

Rayleigh Distribution

$Rayleigh(t; \sigma)$

Let
$$X_1, X_2 \sim Norm(0, \sigma) \quad and \quad Y = \sqrt{X_1^2 + X_2^2}$$
Then
$$Y \sim Rayleigh(\sigma)$$

Chi-square Distribution	Let
$\chi^2(t;v)$	$$X_i \sim Norm(\mu, \sigma^2) \qquad and \qquad Y = \sum_{k=1}^{v} \left(\frac{X_k - \mu}{\sigma}\right)^2$$

Then

$$Y \sim \chi^2(t;v)$$

Binomial Distribution

Limiting Case for constant p:

$$\lim_{\substack{n \to \infty \\ p=p}} Binom(k; n, p) = Norm\big(k; \mu = np, \sigma^2 = np(1-p)\big)$$

$Binom(k; n, p)$

The Normal distribution can be used as an approximation of the Binomial distribution when $np \geq 10$ and $np(1-p) \geq 10$.

$$Binom(k; p, n) \approx Norm\big(t = k + 0.5; \mu = np, \sigma^2 = np(1-p)\big)$$

$$\lim_{\mu \to \infty} F_{Pois}(k; \mu) = F_{Norm}(k; \mu' = \mu, \sigma = \sqrt{\mu})$$

Poisson Distribution

This is a good approximation when $\mu > 1000$. When $\mu > 10$ the same approximation can be made with a correction:

$Pois(k; \mu)$

$$\lim_{\mu \to \infty} F_{Pois}(k; \mu) = F_{Norm}(k; \mu' = \mu - 0.5, \sigma = \sqrt{\mu})$$

For large α and β with fixed α/β:

Beta Distribution

$$Beta(\alpha, \beta) \approx Norm\left(\mu = \frac{\alpha}{\alpha + \beta}, \sigma = \sqrt{\frac{\alpha\beta}{(\alpha + \beta)^2(\alpha + \beta + 1)}}\right)$$

$Beta(t; \alpha, \beta)$

As α and β increase the mean remains constant and the variance is reduced.

Gamma Distribution

Special Case for large k:

$$\lim_{k \to \infty} Gamma(k, \lambda) = Norm\left(\mu = \frac{k}{\lambda}, \sigma = \sqrt{\frac{k}{\lambda^2}}\right)$$

$Gamma(k, \lambda)$

4.6. Pareto Continuous Distribution

Probability Density Function - f(t)

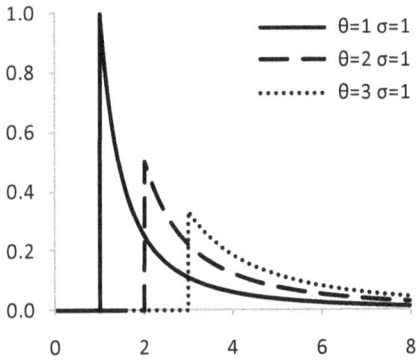

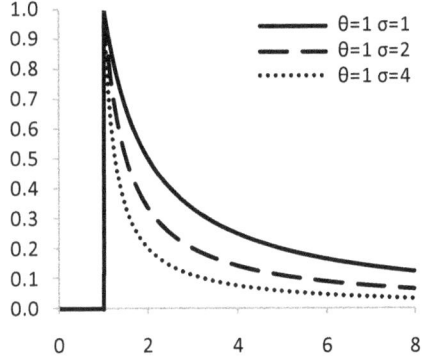

Cumulative Density Function - F(t)

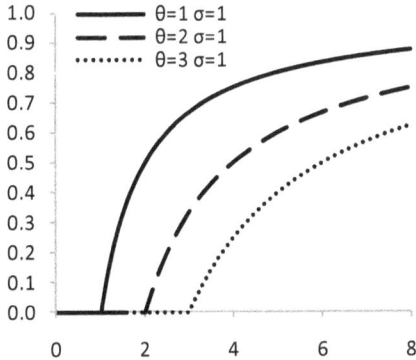

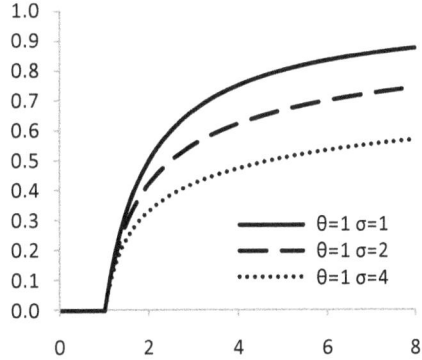

Hazard Rate - h(t)

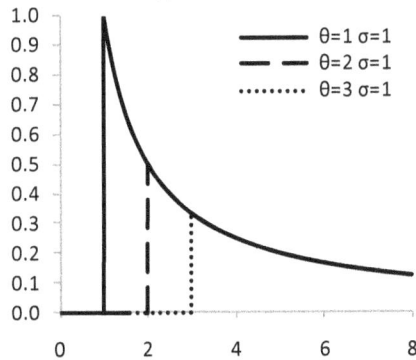

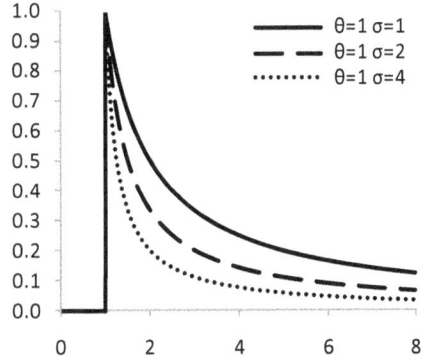

Parameters & Description

Parameters		
	θ	$\theta > 0$	*Location parameter.* θ is the lower limit of t. Sometimes refered to as t-minimum.
	α	$\alpha > 0$	*Shape parameter.* Sometimes called the Pareto index.

Limits	$\theta \le t < \infty$

Distribution	Formulas
PDF	$f(t) = \dfrac{\alpha\theta^\alpha}{t^{\alpha+1}}$
CDF	$F(t) = 1 - \left(\dfrac{\theta}{t}\right)^\alpha$
Reliability	$R(t) = \left(\dfrac{\theta}{t}\right)^\alpha$
Conditional Survivor Function $P(T > x + t \mid T > t)$	$m(x) = R(x\mid t) = \dfrac{R(t+x)}{R(t)} = \dfrac{(t)^\alpha}{(t+x)^\alpha}$ Where, t is the time we know the component has survived to. x is the value of a random variable defined as the time after t. Note: $x = 0$ at t.
Mean Residual Life	$u(t) = \dfrac{\int_t^\infty R(x)dx}{R(t)}$
Hazard Rate	$h(t) = \dfrac{\alpha}{t}$
Cumulative Hazard Rate	$H(t) = \alpha \ln\left(\dfrac{t}{\theta}\right)$

Properties and Moments

Median	$\theta 2^{1/\alpha}$
Mode	θ
Mean - 1st Raw Moment	$\dfrac{\alpha\theta}{\alpha - 1}$, for $\alpha > 1$
Variance - 2nd Central Moment	$\dfrac{\alpha\theta^2}{(\alpha-1)^2(\alpha-2)}$, for $\alpha > 2$
Skewness - 3rd Central Moment	$\dfrac{2(1+\alpha)}{(\alpha-3)}\sqrt{\dfrac{\alpha-2}{\alpha}}$, for $\alpha > 3$

Excess kurtosis - 4th Central Moment	$\dfrac{6(\alpha^3 + \alpha^2 - 6\alpha - 2)}{\alpha(\alpha - 3)(\alpha - 4)}$, for $\alpha > 4$
Characteristic Function	$\alpha(-i\theta t)^{\alpha}\Gamma(-\alpha, -i\theta t)$
100γ % Percentile Function	$t_{\gamma} = \theta(1 - \gamma)^{-1/\alpha}$

Parameter Estimation

Plotting Method

Least Mean Square $y = mx + c$	X-Axis	Y-Axis	$\hat{\alpha} = -m$
	$\ln(t_i)$	$\ln[1 - F]$	$\hat{\theta} = \exp\left\{\dfrac{c}{\hat{\alpha}}\right\}$

Maximum Likelihood Function

Likelihood Function	For complete data:

$$L(\theta, \alpha | E) = \underbrace{\alpha^{n_F}\theta^{\alpha n_F} \prod_{i=1}^{n_F} \frac{1}{t_i^{\alpha+1}}}_{\text{complete failures}}$$

Log-Likelihood Function

$$\Lambda(\theta, \alpha | E) = \underbrace{n_F \ln(\alpha) + n_F\alpha\ln(\theta) - (\alpha + 1)\sum_{i=1}^{n_F} \ln t_i}_{\text{complete failures}}$$

$\dfrac{\partial\Lambda}{\partial\alpha} = 0$ solve for α to get $\hat{\alpha}$:

$$\frac{\partial\Lambda}{\partial\alpha} = \underbrace{-\frac{n_F}{\alpha} + n_F \ln\theta - \sum_{i=1}^{n_F} \ln t_i = 0}_{\text{complete failures}}$$

MLE Point Estimates   The likelihood function increases as θ increases. Therefore, the MLE point estimate is the largest θ which satisfies $\theta \leq t_i < \infty$:

$$\hat{\theta} = \min\{t_1, \dots, t_{n_F}\}$$

Substituting $\hat{\theta}$ gives the MLE for $\hat{\alpha}$:

$$\hat{\alpha} = \frac{n_F}{\sum_{i=1}^{n_F}\left(\ln t_i - \ln\hat{\theta}\right)}$$

Fisher Information

$$I(\theta, \alpha) = \begin{bmatrix} -1/\alpha^2 & 0 \\ 0 & 1/\theta^2 \end{bmatrix}$$

$100\gamma\%$ Confidence Intervals	$\hat{\alpha}$ if θ is unknown	1-Sided Lower	2-Sided Lower	2-Sided Upper
		$\dfrac{\hat{\alpha}}{2n_F}\chi^2_{(1-\gamma)}(2n-2)$	$\dfrac{\hat{\alpha}}{2n_F}\chi^2_{\left\{\frac{1-\gamma}{2}\right\}}(2n-2)$	$\dfrac{\hat{\alpha}}{2n_F}\chi^2_{\left\{\frac{1+\gamma}{2}\right\}}(2n-2)$

| (for complete data) | $\hat{\alpha}$ if θ is known | $\dfrac{\hat{\alpha}}{2n_F}\chi^2_{\{1-\gamma\}}(2n)$ | $\dfrac{\hat{\alpha}}{2n_F}\chi^2_{\{\frac{1-\gamma}{2}\}}(2n)$ | $\dfrac{\hat{\alpha}}{2n_F}\chi^2_{\{\frac{1+\gamma}{2}\}}(2n-2)$ |

(Johnson et al. 1994, p.583) Where $\chi^2_\gamma(n)$ is the $100\gamma^{th}$ percentile of the χ^2-distribution with n degrees of freedom.

Bayesian

Non-informative Priors when θ is known, $\pi_0(\alpha)$
(Yang and Berger 1998, p.22)

Type	Prior
Jeffery and Reference	$\dfrac{1}{\alpha}$

Conjugate Priors

UOI	Likelihood Model	Evidence	Dist. of UOI	Prior Para	Posterior Parameters
b from $Unif(t;a,b)$	Uniform with known a	n_F failures at times t_i	Pareto	θ_0, α_0	$\theta = \max\{t_1,...,t_{n_F}\}$ $\alpha = \alpha_0 + n_F$
θ from $Pareto(t;\theta,\alpha)$	Pareto with known α	n_F failures at times t_i	Pareto	a_0, Θ_0	$a = a_0 - \alpha n_F$ where $a_0 > \alpha n_F$ $\theta = \Theta_0$
α from $Pareto(t;\theta,\alpha)$	Pareto with known θ	n_F failures at times t_i	Gamma	k_0, λ_0	$k = k_0 + n_F$ $\lambda = \lambda_0 + \sum_{i=1}^{n_F}\ln\left(\dfrac{x_i}{\theta}\right)$

Description, Limitations and Uses

Example

5 components are put on a test with the following failure times: 108, 125, 458, 893, 13437 hours

The MLE Estimates are: $\hat{\theta} = 108$

Substituting $\hat{\theta}$ gives the MLE for $\hat{\alpha}$:

$$\hat{\alpha} = \frac{5}{\sum_{i=1}^{n_F}(\ln t_i - \ln(108))} = 0.8029$$

90% confidence interval for $\hat{\alpha}$:

$$\left[\frac{\hat{\alpha}}{10}\chi^2_{\{0.05\}}(8), \quad \frac{\hat{\alpha}}{10}\chi^2_{\{0.95\}}(8)\right]$$
$$[0.2194, \quad 1.2451]$$

Characteristics

80/20 Rule. Most commonly described as the basis for the "80/20 rule" (In a quality context, for example, 80% of manufacturing defects will be a result from 20% of the causes).

Conditional Distribution. The conditional probability distribution given that the event is greater than or equal to a value θ_1 exceeding θ is a Pareto distribution with the same index α but with a minimum θ_1 instead of θ.

Types. This distribution is known as a Pareto distribution of the first kind. The Pareto distribution of the second kind (not detailed here) is also known as the Lomax distribution. Pareto also proposed a third distribution now known as a Pareto distribution of the third kind.

Pareto and the Lognormal Distribution. The Lognormal distribution can model similar physical phenomena as the Pareto distribution. The two distributions have different weights at the extremities.

Let:

$$X_i \sim Pareto(\theta, \alpha_i)$$

Minimum property

$$\min\{X, X_2, \ldots, X_n\} \sim Pareto\left(\theta, \sum_{i=1}^{n} \alpha_i\right)$$

For constant θ.

Applications

Rare Events. The survival function 'slowly' decreases compared to most life distributions which makes it suitable for modeling rare events which have large outcomes. Examples include natural events such as the distribution of the daily rain fall, or the size of manufacturing defects.

Resources

Online:
http://mathworld.wolfram.com/ParetoDistribution.html
http://en.wikipedia.org/wiki/Pareto_distribution
http://socr.ucla.edu/htmls/SOCR_Distributions.html (web calc)

Books:
Arnold, B., 1983. *Pareto distributions*, Fairland, MD: International Co-operative Pub. House.

Johnson, N.L., Kotz, S. & Balakrishnan, N., 1994. *Continuous Univariate Distributions*, Vol. 1 2nd ed., Wiley-Interscience.

Relationship to Other Distributions

Exponential
Distribution
$Exp(t; \lambda)$

Let
$$Y \sim Pareto(\theta, \alpha) \quad and \quad X = \ln(Y/\theta)$$
Then
$$X \sim Exp(\lambda = \alpha)$$

Chi-Squared
Distribution
$\chi^2(x; v)$

Let
$$Y \sim Pareto(\theta, \alpha) \quad and \quad X = 2\alpha \ln(Y/\theta)$$
Then
$$X \sim \chi^2(v = 2)$$
(Johnson et al. 1994, p.526)

Logistic
Distribution
$Logistic(\mu, s)$

Let
$$X \sim Pareto(\theta, \alpha) \quad and \quad Y = -\ln\left\{\left(\frac{X}{\theta}\right)^\alpha - 1\right\}$$
Then
$$Y \sim Logistic(0,1)$$
(Hastings et al. 2000, p.127)

4.7. Triangle Continuous Distribution

Probability Density Function - f(t)

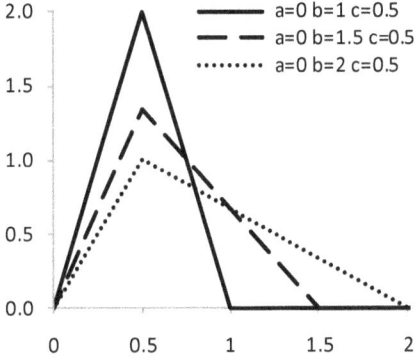

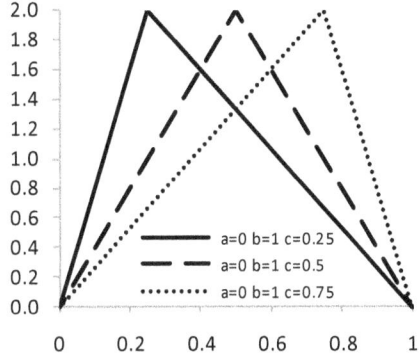

Cumulative Density Function - F(t)

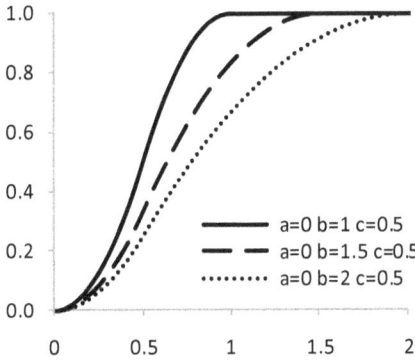

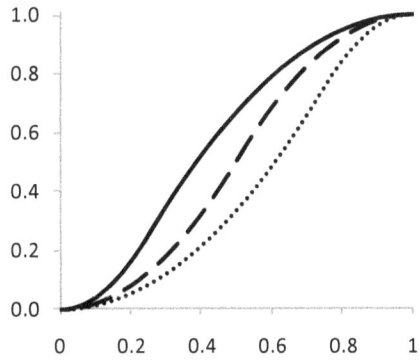

Hazard Rate - h(t)

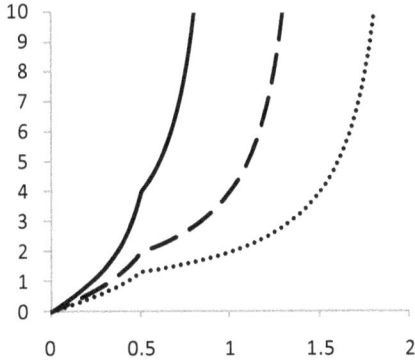

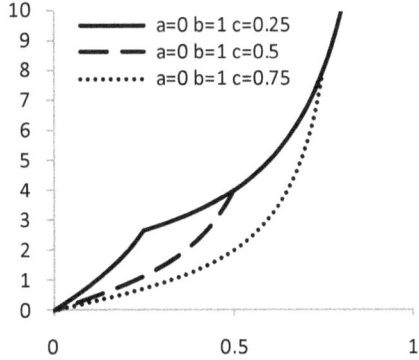

Parameters & Description

Parameters			
	a	$-\infty \leq a < b$	*Minimum Value.* a is the lower bound
	b	$a < b < \infty$	*Maximum Value.* b is the upper bound.
	c	$a \leq c \leq b$	*Mode Value.* c is the mode of the distribution (i.e., apex of the triangle).
Random Variable			$a \leq t \leq b$

Distribution — Formulas

PDF

$$f(t) = \begin{cases} \dfrac{2(t-a)}{(b-a)(c-a)} & \text{for } a \leq t \leq c \\[2ex] \dfrac{2(b-t)}{(b-a)(b-c)} & \text{for } c \leq t \leq b \end{cases}$$

CDF

$$F(t) = \begin{cases} \dfrac{(t-a)^2}{(b-a)(c-a)} & \text{for } a \leq t \leq c \\[2ex] 1 - \dfrac{(b-t)^2}{(b-a)(b-c)} & \text{for } c \leq t \leq b \end{cases}$$

Reliability

$$R(t) = \begin{cases} 1 - \dfrac{(t-a)^2}{(b-a)(c-a)} & \text{for } a \leq t \leq c \\[2ex] \dfrac{(b-t)^2}{(b-a)(b-c)} & \text{for } c \leq t \leq b \end{cases}$$

Properties and Moments

Median

$$a + \sqrt{\tfrac{1}{2}(b-a)(c-a)} \ \ for \ c \geq \frac{b-a}{2}$$
$$b - \sqrt{\tfrac{1}{2}(b-a)(b-c)} \ \ for \ c < \frac{b-a}{2}$$

Mode

$$c$$

Mean - 1st Raw Moment

$$\frac{a+b+c}{3}$$

Variance - 2nd Central Moment

$$\frac{a^2 + b^2 + c^2 - ab - ac - bc}{18}$$

Skewness - 3rd Central Moment

$$\frac{\sqrt{2}(a+b-2c)(2a-b-c)(a-2b+c)}{5(a^2+b^2+c^2-ab-ac-bc)^{3/2}}$$

Excess kurtosis - 4th Central Moment

$$\frac{-3}{5}$$

Characteristic Function

$$-2\frac{(b-c)e^{ita} - (b-a)e^{itc} + (c-a)e^{itb}}{(b-a)(c-a)(b-c)t^2}$$

100γ % Percentile Function

$$t_\gamma = a + \sqrt{\gamma(b-a)(c-a)} \ \ for \ \gamma < F(c)$$
$$t_\gamma = b - \sqrt{(1-\gamma)(b-a)(b-c)} \ \ for \ \gamma \geq F(c)$$

Parameter Estimation

Maximum Likelihood Function

Likelihood Functions

$$L(a,b,c|E) = \underbrace{\prod_{i=1}^{r} \frac{2(t_i - a)}{(b-a)(c-a)}}_{\text{failers to the left of c}} \cdot \underbrace{\prod_{i=r+1}^{n_F} \frac{2(b-t_i)}{(b-a)(b-c)}}_{\text{failures to the right of c}}$$

$$= \left(\frac{2}{b-a}\right)^{n_F} \prod_{i=1}^{r} \frac{t_i - a}{(c-a)} \prod_{i=r+1}^{n_F} \frac{b-t_i}{(b-c)}$$

where failure times are ordered: $T_1 \le T_2 \le \cdots \le T_r \le \cdots \le T_{n_F}$
and r is the number of failure times less than c and s is the number of failure times greater than c. Therefore $n_F = r + s$.

Point Estimates

The MLE estimates $\hat{a}, \hat{b},$ and $\hat{c}$ are obtained by numerically calculating the likelihood function for different r values and selecting the maximum where $\hat{c} = X_{\hat{r}}$.

$$\max_{a \le c \le b} L(a,b,c|E) = \left(\frac{2}{b-a}\right)^{n_F} \{M(a,b,\hat{r}(a,b))\}$$

where

$$M(a,b,r) = \prod_{i=1}^{r-1} \frac{t_i - a}{(t_r - a)} \prod_{i=r+1}^{n_F} \frac{b-t_i}{(b-t_r)}$$

$$r(a,b) = \arg\max_{r\in\{1,\dots,n_F\}} M(a,b,r)$$

Note that the MLE estimates for a and b are not the same as the uniform distribution:

$$\hat{a} \ne \min(t_1^F, t_2^F \dots)$$
$$\hat{b} \ne \max(t_1^F, t_2^F \dots)$$

(Kotz & Dorp 2004)

Description, Limitations and Uses

Example

When eliciting an opinion from an expert on the possible value of a quantity, x, the expert may give:
- Lowest possible value = 0
- Highest possible value = 1
- Estimate of most likely value (mode) = 0.7

The corresponding distribution for x may be a triangle distribution with parameters:
$$a = 0, \qquad b = 1, \qquad c = 0.7$$

Characteristics

Standard Triangle Distribution. The standard triangle distribution has $a = 0, b = 1$. This distribution has a mean at $\sqrt{c/2}$ and median at $1 - \sqrt{(1-c)/2}$.

Symmetrical Triangle Distribution. The symmetrical triangle distribution occurs when $c = (b-a)/2$. The symmetrical triangle distribution is formed from the average of two uniform random variables (see related distributions).

Applications

Subjective Representation. The triangle distribution is often used to model subjective evidence where a and b are the bounds of the estimation and c is an estimation of the mode.

Substitution to the Beta Distribution. Due to the triangle distribution having bounded support it may be used in place of the beta distribution.

Monte Carlo Simulation. Used to approximate distributions of variables when the underlying distribution is unknown. A distribution of interest is obtained by conducting Monte Carlo simulation of a model using the triangle distributions as inputs.

Resources

Online:
http://mathworld.wolfram.com/TriangularDistribution.html
http://en.wikipedia.org/wiki/Triangular_distribution

Books:
Kotz, S. & Dorp, J.R.V., 2004. *Beyond Beta: Other Continuous Families Of Distributions With Bounded Support And Applications*, World Scientific Publishing Company.

Relationship to Other Distributions

Uniform Distribution

$Unif(t; a, b)$

Let

$$X_i \sim Unif(a, b) \qquad and \qquad Y = \frac{X_1 + X_2}{2}$$

Then

$$Y \sim Triangle\left(a, \frac{b-a}{2}, b\right)$$

Beta Distribution

$Beta(t; \alpha, \beta)$

Special Cases:

$$Beta(1,2) = Triangle(0,0,1)$$
$$Beta(2,1) = Triangle(0,1,1)$$

4.8. Truncated Normal Continuous Distribution

Probability Density Function - f(t)

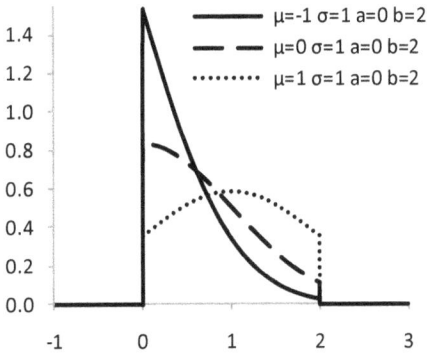

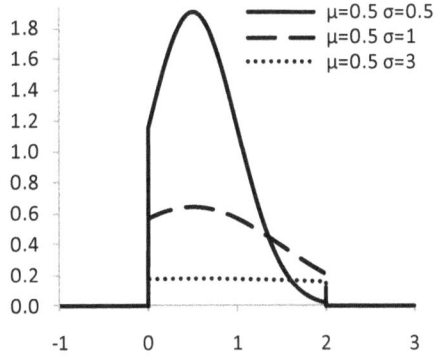

Cumulative Density Function - F(t)

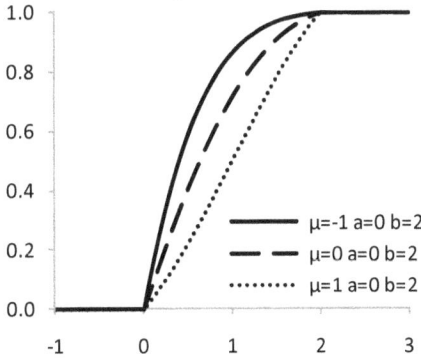

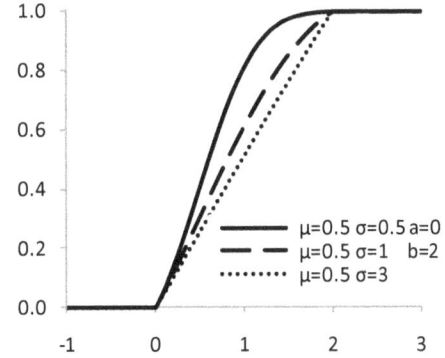

Hazard Rate - h(t)

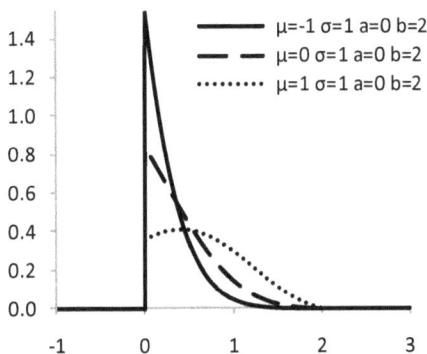

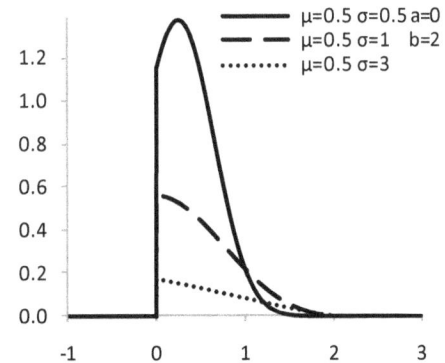

Parameters & Description

Parameters			
	μ	$-\infty < \mu < \infty$	*Location parameter:* The mean of the distribution.
	σ^2	$\sigma^2 > 0$	*Scale parameter:* The standard deviation of the distribution.
	a_L	$-\infty < a_L < b_U$	*Lower Bound:* a_L is the lower bound. The standard normal transform of a_L is $z_a = \frac{a_L - \mu}{\sigma}$.
	b_U	$a_L < b_U < \infty$	*Upper Bound:* b_U is the upper bound. The standard normal transform of b_U is $z_b = \frac{b_U - \mu}{\sigma}$.

Limits	$a_L \leq x \leq b_U$

Distribution	Left Truncated Normal $x \in [0, \infty)$	General Truncated Normal $x \in [a_L, b_U]$
PDF	for $0 \leq x \leq \infty$ $$f(x) = \frac{\phi(z_x)}{\sigma \Phi(-z_0)}$$ otherwise $$f(x) = 0$$	for $a_L \leq x \leq b_U$ $$f(x) = \frac{\frac{1}{\sigma}\phi(z_x)}{\Phi(z_b) - \Phi(z_a)}$$ otherwise $$f(x) = 0$$

where
ϕ is the standard normal pdf with $\mu = 0$ and $\sigma^2 = 1$
Φ is the standard normal cdf with $\mu = 0$ and $\sigma^2 = 1$
$z_i = \left(\frac{i-\mu}{\sigma}\right)$

	Left Truncated Normal	General Truncated Normal
CDF	for $x < 0$ $$F(x) = 0$$ for $0 \leq x < \infty$ $$F(x) = \frac{\Phi(z_x) - \Phi(z_0)}{\Phi(-z_0)}$$	for $x < a_L$ $$F(x) = 0$$ for $a_L \leq x \leq b_U$ $$F(x) = \frac{\Phi(z_x) - \Phi(z_a)}{\Phi(z_b) - \Phi(z_a)}$$ for $x > b_U$ $$F(x) = 1$$
Reliability	for $x < 0$ $$R(x) = 1$$ for $0 \leq x < \infty$ $$R(x) = \frac{\Phi(z_0) - \Phi(z_x)}{\Phi(-z_0)}$$ for $t < 0$	for $x < a_L$ $$R(x) = 1$$ for $a_L \leq x \leq b_U$ $$R(x) = \frac{\Phi(z_b) - \Phi(z_x)}{\Phi(z_b) - \Phi(z_a)}$$ for $x > b_U$ $$R(x) = 0$$ for $t < a_L$

	$m(x) = R(t + x)$	$m(x) = R(t + x)$

Conditional Survivor Function
$P(T > x + t \mid T > t)$

for $0 \le t < \infty$

$$m(x) = R(x \mid t) = \frac{R(t + x)}{R(t)}$$

$$= \frac{1 - \Phi(z_{t+x})}{1 - \Phi(z_t)}$$

$$= \frac{\Phi\left(\frac{\mu - x - t}{\sigma}\right)}{\Phi\left(\frac{\mu - t}{\sigma}\right)}$$

for $a_L \le t \le b_U$

$$m(x) = R(x \mid t) = \frac{R(t + x)}{R(t)}$$

$$= \frac{\Phi(z_b) - \Phi(z_{t+x})}{\Phi(z_b) - \Phi(z_t)}$$

for $t > b_U$

$$m(x) = 0$$

t is the given time we know the component has survived to.
x is the value of a random variable defined as the time after t.
Note: $x = 0$ at t. This operation is the equivalent of t replacing the lower bound.

Mean Residual Life

$$u(t) = \frac{\int_t^\infty R(x)dx}{R(t)} = \frac{\int_t^\infty R(x)dx}{R(t)}$$

Hazard Rate

for $x < 0$
$$h(x) = 0$$

for $0 \le x < \infty$

$$h(x) = \frac{\frac{1}{\sigma}\phi(z_x)[1 - \Phi(z_x)]}{[1 - \Phi(z_0)]^2}$$

for $x < a_L$
$$h(x) = 0$$

for $a_L \le x \le b_U$

$$h(x) = \frac{\frac{1}{\sigma}\phi(z_x)[\Phi(z_b) - \Phi(z_x)]}{[\Phi(z_b) - \Phi(z_a)]^2}$$

for $x > b_U$
$$h(x) = 0$$

Cumulative Hazard Rate

$$H(t) = -\ln[R(t)]$$

$$H(t) = -\ln[R(t)]$$

Properties and Moments	Left Truncated Normal $x \in [0, \infty)$	General Truncated Normal $x \in [a_L, b_U]$
Median	No closed form	No closed form
Mode	μ where $\mu \ge 0$ 0 where $\mu < 0$	μ where $\mu \in [a_L, b_U]$ a_L where $\mu < a_L$ b_U where $\mu > b_U$
Mean 1st Raw Moment	$\mu + \dfrac{\sigma\phi(z_0)}{\Phi(-z_0)}$ where $z_0 = \dfrac{-\mu}{\sigma}$	$\mu + \sigma\dfrac{\phi(z_a) - \phi(z_b)}{\Phi(z_b) - \Phi(z_a)}$ where $z_a = \dfrac{a_L - \mu}{\sigma}$, $\quad z_b = \dfrac{b_U - \mu}{\sigma}$
Variance 2nd Central Moment	$\sigma^2[1 - \{-\Delta_0\}^2 - \Delta_1]$ where $\Delta_k = \dfrac{z_0^k \phi(z_0)}{\Phi(z_0) - 1}$	$\sigma^2[1 - \{-\Delta_0\}^2 - \Delta_1]$ where $\Delta_k = \dfrac{z_b^k \phi(z_b) - z_a^k \phi(z_a)}{\Phi(z_b) - \Phi(z_a)}$

Skewness 3rd Central Moment	$$\frac{-1}{V^{\frac{3}{2}}}[2\Delta_0^3 + (3\Delta_1 - 1)\Delta_0 + \Delta_2]$$

where

$$V = 1 - \Delta_1 - \Delta_0^2$$

Excess kurtosis 4th Central Moment	$$\frac{1}{V^2}[-3\Delta_0^4 - 6\Delta_1\Delta_0^2 - 2\Delta_0^2 - 4\Delta_2\Delta_0 - 3\Delta_1 - \Delta_3 + 3]$$
Characteristic Function	See (Abadir & Magdalinos 2002, pp.1276-1287)
100α% Percentile Function	$t_\alpha =$ $\mu + \sigma\Phi^{-1}\{\alpha + \Phi(z_0)[1-\alpha]\}$ $t_\alpha =$ $\mu + \sigma\Phi^{-1}\{\alpha\Phi(z_b) + \Phi(z_a)[1-\alpha]\}$

Parameter Estimation

Maximum Likelihood Function

Likelihood
Function

For limits $[a_L, b_U]$:

$$L(\mu, \sigma, a_L, b_U) = \frac{1}{\left(\sigma\sqrt{2\pi}\{\Phi(z_b) - \Phi(z_a)\}\right)^{n_F}} \underbrace{\prod_{i=1}^{n_F} \exp\left(-\frac{1}{2}\left[\frac{x_i - \mu}{\sigma}\right]^2\right)}_{\text{complete failures}}$$

$$= \frac{1}{\left(\sigma\sqrt{2\pi}\{\Phi(z_b) - \Phi(z_a)\}\right)^{n_F}} \underbrace{\exp\left(-\frac{1}{2\sigma^2}\sum_{i=1}^{n_F}(x_i - \mu)^2\right)}_{\text{complete failures}}$$

For limits $[0, \infty)$

$$L(\mu, \sigma) = \frac{1}{\left(\Phi\{-z_0\}\sigma\sqrt{2\pi}\right)^{n_F}} \underbrace{\prod_{i=1}^{n_F} \exp\left(-\frac{1}{2}\left[\frac{x_i - \mu}{\sigma}\right]^2\right)}_{\text{complete failures}}$$

$$= \frac{1}{\left(\Phi\{-z_0\}\sigma\sqrt{2\pi}\right)^{n_F}} \underbrace{\exp\left(-\frac{1}{2\sigma^2}\sum_{i=1}^{n_F}(x_i - \mu)^2\right)}_{\text{complete failures}}$$

Log-Likelihood
Function

For limits $[a_L, b_U]$:

$$\Lambda(\mu, \sigma, a_L, b_U | E)$$

$$= -n_F \ln[\Phi(z_b) - \Phi(z_a)] - n_F \ln\left(\sigma\sqrt{2\pi}\right) - \underbrace{\frac{1}{2\sigma^2}\sum_{i=1}^{n_F}(x_i - \mu)^2}_{\text{complete failures}}$$

For limits $[0, \infty)$

$$\Lambda(\mu, \sigma | E) = -n_F \ln(\Phi\{-z_0\}) - n_F \ln\left(\sigma\sqrt{2\pi}\right) - \underbrace{\frac{1}{2\sigma^2}\sum_{i=1}^{n_F}(x_i - \mu)^2}_{\text{complete failures}}$$

$$\frac{\partial \Lambda}{\partial \mu} = 0$$

$$\underbrace{\frac{\partial \Lambda}{\partial \mu} = \frac{-n_F}{\sigma}\left[\frac{\phi(z_a) - \phi(z_b)}{\Phi(z_b) - \Phi(z_a)}\right] + \frac{1}{\sigma^2}\sum_{i=1}^{n_F}(x_i - \mu) = 0}_{\text{complete failures}}$$

$$\frac{\partial \Lambda}{\partial \sigma} = 0$$

$$\underbrace{\frac{\partial \Lambda}{\partial \sigma} = \frac{-n_F}{\sigma^2}\left[\frac{z_a\phi(z_a) - z_b\phi(z_b)}{\Phi(z_b) - \Phi(z_a)}\right] - \frac{n_F}{\sigma} + \frac{1}{\sigma^3}\sum_{i=1}^{n_F}(x_i - \mu)^2 = 0}_{\text{complete failures}}$$

MLE Point Estimates

First Estimate the values for z_a and z_b by solving the simultaneous equations numerically (Cohen 1991, p.33):

$$H_1(z_a, z_b) = \frac{Q_a - Q_b - z_a}{z_b - z_a} = \frac{\bar{x} - a_L}{b_U - a_L}$$

$$H_2(z_a, z_b) = \frac{1 + z_a Q_a - z_b Q_b - (Q_a - Q_b)^2}{(z_b - z_a)^2} = \frac{s^2}{(b_U - a_L)^2}$$

Where:

$$Q_a = \frac{\phi(z_a)}{\Phi(z_b) - \Phi(z_a)}, \quad Q_b = \frac{\phi(z_b)}{\Phi(z_b) - \Phi(z_a)}$$

$$z_a = \frac{a_L - \mu}{\sigma}, \qquad z_b = \frac{b_U - \mu}{\sigma}$$

$$\bar{x} = \frac{1}{n^F}\sum_{0}^{n_F} x_i, \quad s^2 = \frac{1}{n_F - 1}\sum_{0}^{n_F}(x_i - \bar{x})^2$$

The distribution parameters can then be estimated using:

$$\hat{\sigma} = \frac{b_U - a_L}{\widehat{z_b} - \widehat{z_a}}, \qquad \hat{\mu} = a_L - \hat{\sigma}\widehat{z_a}$$

(Cohen 1991, p.44) provides a graphical procedure to estimate parameters to use as the starting point for numerical solvers.

For the case where the limits are $[0, \infty)$ first numerically solve for z_0:

$$\frac{1 - Q_0(Q_0 - z_0)}{(Q_0 - z_0)^2} = \frac{s^2}{\bar{x}}$$

where

$$Q_0 = \frac{\phi(z_0)}{1 - \Phi(z_0)}$$

The distribution parameters can be estimated using:

$$\hat{\sigma} = \frac{\bar{x}}{Q_0 - \widehat{z_0}}, \qquad \hat{\mu} = -\hat{\sigma}\widehat{z_0}$$

When the limits a_L and b_U are unknown, the likelihood function is maximized when the difference, $\Phi(z_b) - \Phi(z_a)$, is at its minimum. This occurs when the difference between $b_U - a_L$ is at its minimum. Therefore, the MLE estimates for a_L and b_U are:

$$\widehat{a_L} = \min(t_1^F, t_2^F \ldots)$$
$$\widehat{b_U} = \max(t_1^F, t_2^F \ldots)$$

Fisher Information (Cohen 1991, p.40)	$I(\mu, \sigma^2) = \begin{bmatrix} \frac{1}{\sigma^2}[1 - Q'_a + Q'_b] & \frac{1}{\sigma^2}\left[\frac{2(\bar{x} - \mu)}{\sigma} - \lambda_a + \lambda_b\right] \\ \frac{1}{\sigma^2}\left[\frac{2(\bar{x} - \mu)}{\sigma} - \lambda_a + \lambda_b\right] & \frac{1}{\sigma^2}\left[\frac{3[s^2 + (\bar{x} - \mu)^2]}{\sigma^2} - 1 - \eta_a + \eta_b\right] \end{bmatrix}$

Where

$$Q'_a = Q_a(Q_a - z_a), \qquad Q'_b = -Q_b(Q_b + z_b)$$
$$\lambda_a = a_L Q'_a + Q_a, \qquad \lambda_b = b_U Q'_b + Q_b$$
$$\eta_a = a_L(\lambda_a + Q_a), \qquad \eta_b = b_U(\lambda_b + Q_b)$$

$100\gamma\%$ Confidence Intervals	Calculated from the Fisher information matrix. See Section 1.4.7. For further detail and examples see (Cohen 1991, p.41)

Bayesian

No closed form solutions to priors exist.

Description, Limitations and Uses

Example 1	The size of washers delivered from a manufacturer is desired to be modeled. The manufacture has already removed all washers below 15.95mm and washers above 16.05mm. The washers received have the following diameters:

$$15.976,\ 15.970,\ 15.955,\ 16.007,\ 15.966,\ 15.952,\ 15.955\ mm$$

From data:
$$\bar{x} = 15.973, \qquad s^2 = 4.3950E\text{-}4$$

Using numerical solver MLE Estimates for z_a and z_b are:

$$\widehat{z_a} = 0, \qquad \widehat{z_b} = 3.3351$$

Therefore

$$\hat{\sigma} = \frac{b_U - a_L}{\widehat{z_b} - \widehat{z_a}} = 0.029984$$

$$\hat{\mu} = a_L - \hat{\sigma}\widehat{z_a} = 15.95$$

To calculate confidence intervals, first calculate:
$$Q'_a = 0.63771, \qquad Q'_b = -0.010246$$
$$\lambda_a = 10.970, \qquad \lambda_b = -0.16138$$
$$\eta_a = 187.71, \qquad \eta_b = -2.54087$$

90% confidence intervals:
$$I(\mu, \sigma) = \begin{bmatrix} 391.57 & -10699 \\ -10699 & -209183 \end{bmatrix}$$

$$[J_n(\hat{\mu}, \hat{\sigma})]^{-1} = [n_F I(\hat{\mu}, \hat{\sigma})]^{-1} = \begin{bmatrix} 1.1835E\text{-}4 & -6.0535E\text{-}6 \\ -6.0535E\text{-}6 & -2.2154E\text{-}7 \end{bmatrix}$$

90% confidence interval for μ:
$$\left[\hat{\mu} - \Phi^{-1}(0.95)\sqrt{1.1835E\text{-}4}, \qquad \hat{\mu} + \Phi^{-1}(0.95)\sqrt{1.1835E\text{-}4}\right]$$
$$[15.932,\ 15.968]$$

90% confidence interval for σ:

$$\left[\hat{\sigma} \cdot \exp \left\{ \frac{\Phi^{-1}(0.95)\sqrt{2.2154E\text{-}7}}{-\hat{\sigma}} \right\}, \quad \hat{\sigma} \cdot \exp \left\{ \frac{\Phi^{-1}(0.95)\sqrt{2.2154E\text{-}7}}{\hat{\sigma}} \right\} \right]$$

$$[2.922E\text{-}2, \quad 3.0769E\text{-}2]$$

An estimate can be made on how many washers the manufacturer discards:

The distribution of washer sizes is a Normal Distribution with estimated parameters $\hat{\mu} = 15.95$, $\hat{\sigma} = 0.029984$. The percentage of washers wish pass quality control is:

$$F(16.05) - F(15.95) = 49.96\%$$

It is likely that there is too much variance in the manufacturing process for this system to be efficient.

Example 2   The following example adjusts the calculations used in the Normal Distribution to account for the fact that the limit on distance is $[0, \infty)$.

The accuracy of a cutting machine used in manufacturing is desired to be measured. 5 cuts at the required length are made and measured as:
$$7.436, \ 10.270, \ 10.466, \ 11.039, \ 11.854 \ mm$$

From data:
$$\bar{x} = 10.213, \qquad s^2 = 2.789$$

Using numerical solver MLE Estimates for z_0 is:

$$\hat{z_0} = -4.5062$$

Therefore

$$\hat{\sigma} = \frac{\bar{x}}{Q_0 - \hat{z_0}} = 2.26643$$

$$\hat{\mu} = -\hat{\sigma}\hat{z_a} = 10.213$$

To calculate confidence intervals, first calculate:
$$Q_0' = 7.0042E\text{-}5, \qquad \lambda_0 = 1.5543E\text{-}5, \qquad \lambda_b = -0.16138$$

90% confidence intervals:
$$I(\mu, \sigma) = \begin{bmatrix} 0.19466 & -2.9453E\text{-}6 \\ -2.9453E\text{-}6 & 0.12237 \end{bmatrix}$$

$$[J_n(\hat{\mu}, \hat{\sigma})]^{-1} = [n_F I(\hat{\mu}, \hat{\sigma})]^{-1} = \begin{bmatrix} 1.0274 & 2.4728E\text{-}5 \\ 2.4728E\text{-}5 & 1.6343 \end{bmatrix}$$

90% confidence interval for μ:
$$\left[\hat{\mu} - \Phi^{-1}(0.95)\sqrt{1.0274}, \quad \hat{\mu} + \Phi^{-1}(0.95)\sqrt{1.0274} \right]$$
$$[8.546, \ 11.88]$$
90% confidence interval for σ:

$$\left[\hat{\sigma}.\exp\left\{\frac{\Phi^{-1}(0.95)\sqrt{1.6343}}{-\hat{\sigma}}\right\}, \quad \hat{\sigma}.\exp\left\{\frac{\Phi^{-1}(0.95)\sqrt{1.6343}}{\hat{\sigma}}\right\}\right]$$

$$[0.8962, \quad 5.732]$$

To compare these results to a non-truncated normal distribution:

	90% Lower CI	Point Est	90% Upper CI
Norm - μ Classical	10.163	10.213	10.262
Norm - σ^2 Classical	1.176	2.789	15.697
Norm - μ Bayesian	8.761	10.213	11.665
Norm - σ^2 Bayesian	0.886	2.789	6.822
TNorm - μ	8.546	10.213	11.88
TNorm - σ^2	0.80317	5.1367	32.856

*Note: The TNorm σ estimate and interval are squared.

The truncated normal produced results which had a wider confidence in the parameter estimates, however the point estimates were within each other confidence intervals. In this case the truncation correction might be ignored for ease of calculation.

Characteristics

For large μ/σ truncation may have negligible affect. In this case the use the Normal Continuous Distribution as an approximation.

Let: $X \sim TNorm(\mu, \sigma^2)$ where $X \in [a, b]$

Convolution Property. The sum of truncated normal distribution random variables is not a truncated normal distribution. When truncation is symmetrical about the mean the sum of truncated normal distribution random variables is well approximated using:

$$Y = \sum_{i=1}^{n} X_i \quad \text{where} \quad \frac{b_i - a_i}{2} = \mu_i$$

$$Y \approx TNorm\left(\sum \mu_i, \sum Var(X_i)\right) \quad \text{where} \quad Y \in [\sum a_i, \sum b_i]$$

Linear Transformation Property (Cozman & Krotkov 1997)

$$Y = cX + d$$
$$Y \sim TNorm(c\mu + d, d^2\sigma^2) \quad \text{where} \quad Y \in [ca + d, cb + d]$$

Applications

Life Distribution. When used as a life distribution a truncated Normal Distribution may be used due to the constraint t≥0. However, it is often found that the difference in results is negligible. (Rausand & Høyland 2004)

Repair Time Distributions. The truncated normal distribution may be used to model simple repair or inspection tasks that have a typical duration with little variation using the limits $[0, \infty)$

Failures After Pre-test Screening. When a customer receives a product from a vendor, the product may have already been subject to burn-in testing. The customer will not know the number of failures which occurred during the burn-in, but may know the duration. As such the failure distribution is left truncated. (Meeker & Escobar 1998, p.269)

Flaws under the inspection threshold.   When a flaw is not detected due to the flaw's amplitude being less than the inspection threshold the distribution is left truncated. (Meeker & Escobar 1998, p.266)

Worst Case Measurements. Sometimes only the worst performers from a population are monitored and have data collected. Therefore, the threshold which determined that the item be monitored is the truncation limit. (Meeker & Escobar 1998, p.267)

Screening Out Units with Large Defects. In quality control processes it may be common to remove defects which exceed a limit. The remaining population of defects delivered to the customer has a right truncated distribution. (Meeker & Escobar 1998, p.270)

Resources

Online:
http://en.wikipedia.org/wiki/Truncated_normal_distribution
http://socr.ucla.edu/htmls/SOCR_Distributions.html (web calc)
http://www.ntrand.com/truncated-normal-distribution/

Books:
Cohen, 1991. *Truncated and Censored Samples* 1st ed., CRC Press.

Patel, J.K. & Read, C.B., 1996. *Handbook of the Normal Distribution* 2nd ed., CRC.

Schneider, H., 1986. *Truncated and censored samples from normal populations*, M. Dekker.

Relationship to Other Distributions

Normal Distribution

$Norm(x; \mu, \sigma^2)$

Let:

$$X \sim Norm(\mu, \sigma^2)$$
$$X \in (\infty, \infty)$$

Then:

$$Y \sim TNorm(\mu, \sigma^2, a_L, b_U)$$
$$Y \in [a_L, b_U]$$

For further relationships see Normal Continuous Distribution

4.9. Uniform Continuous Distribution

Probability Density Function - f(t)

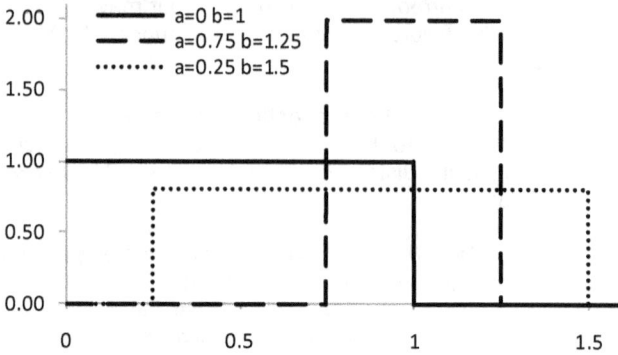

Cumulative Density Function - F(t)

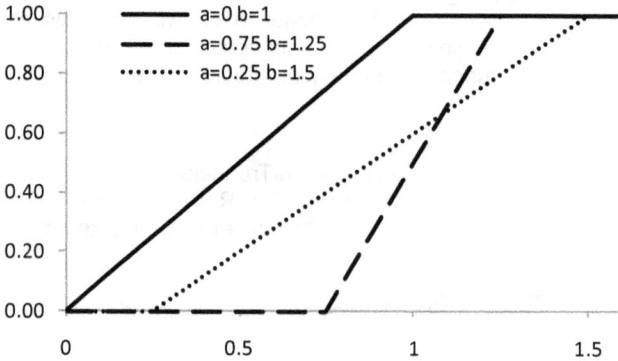

Hazard Rate - h(t)

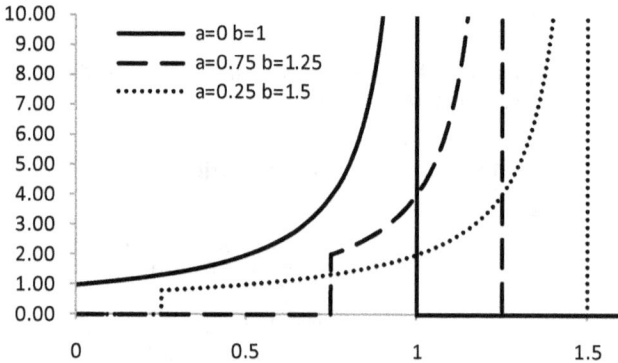

Parameters & Description			
Parameters	a	$0 \le a < b$	*Minimum Value.* a is the lower bound of the uniform distribution.
	b	$a < b < \infty$	*Maximum Value.* b is the upper bound of the uniform distribution.
Random Variable		$a \le t \le b$	

Distribution	Time Domain	Laplace
PDF	$$f(t) = \begin{cases} \dfrac{1}{b-a} & \text{for } a \le t \le b \\ 0 & \text{otherwise} \end{cases}$$ $$= \frac{1}{b-a}\{u(t-a) - u(t-b)\}$$ Where $u(t-a)$ is the Heaviside step function.	$$f(s) = \frac{e^{-as} - e^{-bs}}{s(b-a)}$$
CDF	$$F(t) = \begin{cases} 0 & \text{for } t < a \\ \dfrac{t-a}{b-a} & \text{for } a \le t \le b \\ 1 & \text{for } t > b \end{cases}$$ $$= \frac{t-a}{b-a}\{u(t-a) - u(t-b)\}$$ $$+ u(t-b)$$	$$F(s) = \frac{e^{-as} - e^{-bs}}{s^2(b-a)}$$
Reliability	$$R(t) = \begin{cases} 1 & \text{for } t < a \\ \dfrac{b-t}{b-a} & \text{for } a \le t \le b \\ 0 & \text{for } t > b \end{cases}$$	$$R(s) = \frac{e^{-bs} - e^{-as}}{s^2(b-a)} + \frac{1}{s}$$
Conditional Survivor Function $P(T > x + t \mid T > t)$	For $t < a$: $$m(x) = \frac{R(t+x)}{R(t)} = \begin{cases} 1 & \text{for } t + x < a \\ \dfrac{b-(t+x)}{b-a} & \text{for } a \le t + x \le b \\ 0 & \text{for } t > b \end{cases}$$ For $a \le t \le b$: $$m(x) = \frac{R(t+x)}{R(t)} = \begin{cases} 1 & \text{for } t + x < a \\ \dfrac{b-(t+x)}{b-t} & \text{for } a \le t + x \le b \\ 0 & \text{for } t + x > b \end{cases}$$ For $t > b$: $$m(x) = 0$$ Where t is the given time we know the component has survived to. x is value of the random variable defined as the time after t. Note: $x = 0$ at t.	
Mean Residual Life	For $t < a$: $$u(t) = \tfrac{1}{2}(a+b) - t$$ For $a \le t \le b$:	

$$u(t) = a - t - \frac{(a-b)^2}{2(t-b)}$$

For $t > b$:

$$u(t) = 0$$

Hazard Rate

$$h(t) = \begin{cases} \dfrac{1}{b-t} & \text{for } a \leq t \leq b \\ 0 & \text{otherwise} \end{cases}$$

Cumulative
Hazard Rate

$$H(t) = \begin{cases} 0 & \text{for } t < a \\ -\ln\left(\dfrac{b-t}{b-a}\right) & \text{for } a \leq t \leq b \\ \infty & \text{for } t > b \end{cases}$$

Properties and Moments

Median	$\frac{1}{2}(a+b)$
Mode	Any value between a and b
Mean - 1st Raw Moment	$\frac{1}{2}(a+b)$
Variance - 2nd Central Moment	$\frac{1}{12}(b-a)^2$
Skewness - 3rd Central Moment	0
Excess kurtosis - 4th Central Moment	$-\frac{6}{5}$
Characteristic Function	$\dfrac{e^{itb} - e^{ita}}{it(b-a)}$
100α% Percentile Function	$t_\alpha = \alpha(b-a) + a$

Parameter Estimation

Maximum Likelihood Function

Likelihood
Functions

$$L(a,b|E) = \underbrace{\left(\frac{1}{b-a}\right)^{n_F}}_{\text{complete failures}} \cdot \underbrace{\prod_{i=1}^{n_S}\left(\frac{b-t_i^S}{b-a}\right)}_{\text{survivors}} \cdot \underbrace{\prod_{i=1}^{n_I}\left(1 + \frac{t_i^{RI} - t_i^{LI}}{b-a}\right)}_{\text{interval failures}}$$

This assumes that all times are within the bound a, b.

When there is only complete failure data:

$$L(a,b|E) = \left(\frac{1}{b-a}\right)^{n_F}$$

where

$$a \leq t_i \leq b$$

Point
Estimates

The likelihood function is maximized when a is large, b is small with the restriction that all times are between a and b. Thus:

$$\hat{a} = \min(t_1^F, t_2^F \ldots)$$
$$\hat{b} = \max(t_1^F, t_2^F \ldots)$$

When $a = 0$ and b is estimated with complete data the following estimates may be used where $t_{max} = \max(t_1^F, t_2^F \ldots t_n^F)$. (Johnson et al. 1995, p.286)

1. MLE. $\hat{b} = t_{max}$

2. Min Mean Square Error. $\hat{b} = \dfrac{n+2}{n+1} t_{max}$

3. Unbiased Estimator. $\hat{b} = \dfrac{n+1}{n} t_{max}$

4. Closest Estimator. $\hat{b} = 2^{1/n}\, t_{max}$

Procedures for parameter estimating when there is censored data is detailed in (Johnson et al. 1995, p.286)

Fisher
Information

$$I(a,b) = \begin{bmatrix} \dfrac{-1}{(a-b)^2} & \dfrac{1}{(a-b)^2} \\ \dfrac{1}{(a-b)^2} & \dfrac{-1}{(a-b)^2} \end{bmatrix}$$

Bayesian

The Uniform distribution is widely used in Bayesian methods as a non-informative prior or to model evidence which only suggests bounds on the parameter.

Non-informative Prior. The Uniform distribution can be used as a non-informative prior. As can be seen below, the only affect the uniform prior has on Bayes equation is to limit the range of the parameter for which the denominator integrates over.

$$\pi(\theta|E) = \frac{L(E|\theta)\left(\frac{1}{b-a}\right)}{\int_a^b L(E|\theta)\left(\frac{1}{b-a}\right)d\theta} = \frac{L(E|\theta)}{\int_a^b L(E|\theta)\,d\theta}$$

Parameter Bounds. This distribution allows an easy method to mathematically model soft data where only the parameter bounds can be estimated. An example is where uniform distribution can model a person's opinion on the value of θ, where they know that it could not be lower than a or greater than b, but is unsure of any particular value θ could take.

Non-informative Priors

Jeffrey's Prior $\dfrac{1}{a - b}$

Description, Limitations and Uses

Example

For an example of the uniform distribution being used in Bayesian updating as a prior, $Beta(1,1)$ see the binomial distribution.

Given the following data calculate the MLE parameter estimates:

240, 585, 223, 751, 255

$$\hat{a} = 223$$
$$\hat{b} = 751$$

Characteristics	The Uniform distribution is a special case of the Beta distribution when $\alpha = \beta = 1$.

The uniform distribution has an increasing failure rate with $\lim_{t \to b} h(t) = \infty$.

The Standard Uniform Distribution has parameters $a = 0$ and $b = 1$. This results in $f(t) = 1$ for $a \leq t \leq b$ and 0 otherwise.

$$T \sim Unif(a, b)$$

Uniformity Property
If $t > a$ and $t + \Delta < b$ then:

$$P(t \to t + \Delta) = \int_t^{t+\Delta} \frac{1}{b-a} dx = \frac{\Delta}{b-a}$$

The probability that a random variable falls within any interval of fixed length is independent of the location, t, and is only dependent on the interval size, Δ.

Variate Generation Property
$$F^{-1}(u) = u(b - a) + a$$

Residual Property
If k is a real constant where $a < k < b$ then:
$$\Pr(T|T > k) \sim Unif(a = k, b)$$

Applications

Random Number Generator. The uniform distribution is widely used as the basis for the generation of random numbers for other statistical distributions. The random uniform values are mapped to the desired distribution by solving the inverse cdf.

Bayesian Inference. The uniform distribution can be used as a non-informative prior and to model soft evidence.

Special Case of Beta Distribution. In applications like Bayesian statistics the uniform distribution is used as an uninformative prior by using a beta distribution of $\alpha = \beta = 1$.

Resources

Online:
http://mathworld.wolfram.com/UniformDistribution.html
http://en.wikipedia.org/wiki/Uniform_distribution_(continuous)
http://socr.ucla.edu/htmls/SOCR_Distributions.html

Books:
Johnson, N.L., Kotz, S. & Balakrishnan, N., 1995. *Continuous Univariate Distributions*, Vol. 2 2nd ed., Wiley-Interscience.

Relationship to Other Distributions

Beta Distribution	Let
	$$X_i \sim Unif(0,1) \quad and \quad X_1 \leq X_2 \leq \cdots \leq X_n$$
$Beta(t; \alpha, \beta, a, b)$	Then
	$$X_r \sim Beta(r, n - r + 1)$$

Where n and k are integers.

Special Case:
$$Beta(t; a, b | \alpha = 1, \beta = 1) = Unif(t; a, b)$$

Exponential
Distribution

Let
$$X \sim Exp(\lambda) \quad and \quad Y = \exp(-\lambda X)$$
Then

$Exp(t; \lambda)$

$$Y \sim Unif(0,1)$$

5. Univariate Discrete Distributions

5.1. Bernoulli Discrete Distribution

Probability Density Function - f(k)

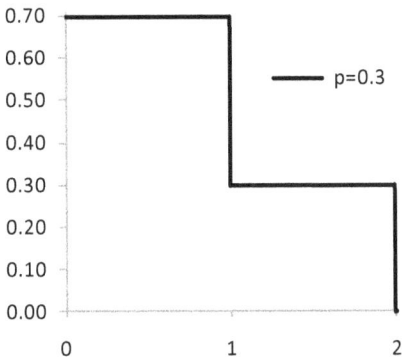

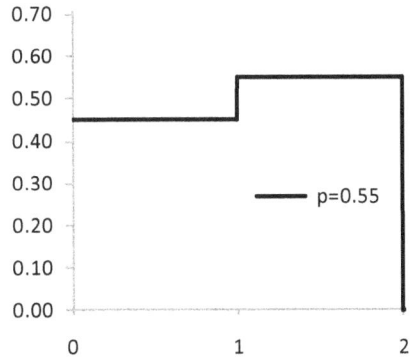

Cumulative Density Function - F(k)

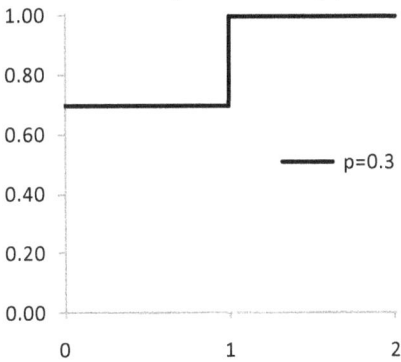

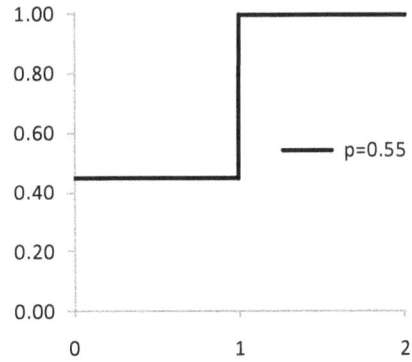

Hazard Rate - h(k)

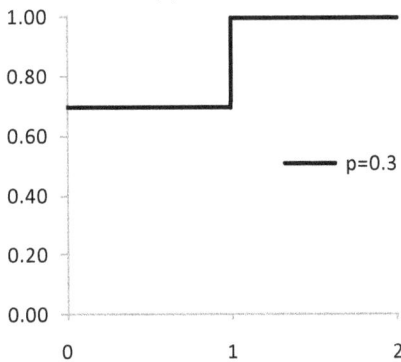

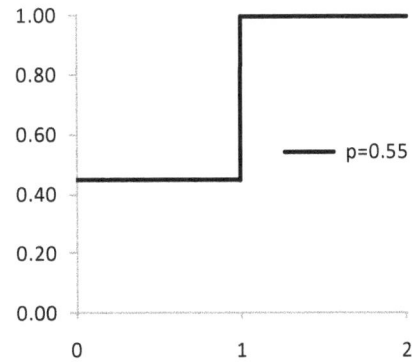

Parameters & Description

Parameters	p	$0 \leq p \leq 1$	Bernoulli probability parameter. Probability of success.
Random Variable		$k \in \{0,1\}$	

Calculates the probability of getting exactly k (0 or 1) successes in 1 trial with probability p.

Distribution	Formulas
PDF	$f(k) = p^k(1-p)^{1-k}$ $= \begin{cases} 1-p & \text{for } k=0 \\ p & \text{for } k=1 \end{cases}$
CDF	$F(k) = (1-p)^{1-k}$ $= \begin{cases} 1-p & \text{for } k=0 \\ 1 & \text{for } k=1 \end{cases}$
Reliability	$R(k) = 1 - (1-p)^{1-k}$ $= \begin{cases} p & \text{for } k=0 \\ 0 & \text{for } k=1 \end{cases}$
Hazard Rate	$h(k) = \begin{cases} 1-p & \text{for } k=0 \\ 1 & \text{for } k=1 \end{cases}$

Properties and Moments

Mode	$k_{0.5} = \|p\|$ when $p \neq 0.5$ $k_{0.5} = \{0,1\}$ when $p = 0.5$
Mean - 1st Raw Moment	p
Variance - 2nd Central Moment	$p(1-p)$
Skewness - 3rd Central Moment	$\dfrac{q-p}{\sqrt{pq}}$ where $q = (1-p)$
Excess kurtosis - 4th Central Moment	$\dfrac{6p^2 - 6p + 1}{p(1-p)}$
Characteristic Function	$(1-p) + pe^{it}$

Parameter Estimation

Maximum Likelihood Function

Likelihood Function
$$L(p|E) = p^{\sum k_i}(1-p)^{n-\sum k_i}$$

where n is the number of Bernoulli trials $k_i \in \{0,1\}$, and $\sum k_i = \sum_{i=1}^{n} k_i$

$\dfrac{dL}{dp} = 0$	solve for p

$$\frac{dL}{dp} = \sum k. \, p^{\sum(k_i)-1}(1-p)^{n-\sum k_i} - (n-\sum k)p^{\sum k_i}(1-p)^{n-1-\sum k_i} = 0$$

$$\sum k. \, p^{\sum(k_i)-1}(1-p)^{n-\sum k_i} = (n-\sum k_i)p^{\sum k_i}(1-p)^{n-1-\sum k_i}$$

$$\sum k_i. \, p^{-1} = (n-\sum k_i)(1-p)^{-1}$$

$$\frac{(1-p)}{p} = \frac{n-\sum k_i}{\sum k_i}$$

$$p = \frac{\sum k_i}{n}$$

Fisher Information	$I(p) = \dfrac{1}{p(1-p)}$

MLE Point Estimates	The MLE point estimate for p:

$$\hat{p} = \frac{\sum k}{n}$$

Fisher Information	$I(p) = \dfrac{1}{p(1-p)}$

Confidence Intervals	See discussion in binomial distribution.

Bayesian

Non-informative Priors for p, $\pi(p)$
(Yang and Berger 1998, p.6)

Type	Prior	Posterior
Uniform Proper Prior with limits $p \in [a,b]$	$\dfrac{1}{b-a}$	Truncated Beta Distribution For $a \le p \le b$ $c.Beta(p; 1+k, 2-k)$ Otherwise, $\pi(p) = 0$
Uniform Improper Prior with limits $p \in [0,1]$	$1 = Beta(p; 1,1)$	$Beta(p; 1+k, 2-k)$
Jeffrey's Prior Reference Prior	$\dfrac{1}{\sqrt{p(1-p)}} = Beta\left(p; \dfrac{1}{2}, \dfrac{1}{2}\right)$	$Beta\left(p; \dfrac{1}{2}+k, 1.5-k\right)$ when $p \in [0,1]$
MDIP	$1.6186p^p(1-p)^{1-p}$	Proper - No Closed Form
Novick and Hall	$p^{-1}(1-p)^{-1} = Beta(0,0)$	$Beta(p; k, 1-k)$ when $p \in [0,1]$

Conjugate Priors

UOI	Likelihood Model	Evidence	Dist. of UOI	Prior Para	Posterior Parameters
p from $Bernoulli(k; p)$	Bernoulli	k failures in 1 trail	Beta	α_0, β_0	$\alpha = \alpha_o + k$ $\beta = \beta_o + 1 - k$

Description, Limitations and Uses

Example

When a demand is placed on a machine it undergoes a Bernoulli trial with success defined as a proper start. It is known the probability of a proper start, p, equals 0.8. Therefore, the probability the machine fails to start: $f(0) = 0.2$.

For an example with multiple Bernoulli trials see the binomial distribution.

Characteristics

A Bernoulli process is a probabilistic experiment that can have one of two outcomes, success ($k = 1$) with the probability of success is p, and failure ($k = 0$) with the probability of failure is $q \equiv 1 - p$.

Single Trial. It's important to emphasis that the Bernoulli distribution is for a single trial or event. The case of multiple Bernoulli trials with replacement is the binomial distribution. The case of multiple Bernoulli trials without replacement is the hypergeometric distribution.

$$K \sim Bernoulli(k|p)$$

Maximum Property

$$\max\{K_1, K_2, \dots, K_n\} \sim Bernoulli(k; p = 1 - \Pi\{1 - p_i\})$$

Minimum property

$$\min\{K_1, K_2, \dots, K_n\} \sim Bernoulli(k; p = \Pi p_i)$$

Product Property

$$\prod_{i=1}^{n} K_i \sim Bernoulli(\Pi k; p = \Pi p_i)$$

Applications

Used to model a single event with only two outcomes. In reliability engineering it is most often used to model demands or shocks to a component, where the component will fail with probability p.

In practice it is rare for only a single event to be considered and so a binomial distribution is most often used (with the assumption of replacement). The conditions and assumptions of a Bernoulli trial however are used as the basis for each trial in a binomial distribution. See 'Related Distributions' and binomial distribution for more details.

Resources

Online:
http://mathworld.wolfram.com/BernoulliDistribution.html
http://en.wikipedia.org/wiki/Bernoulli_distribution
http://socr.ucla.edu/htmls/SOCR_Distributions.html (web calc)

Books:
Collani, E.V. & Dräger, K., 2001. *Binomial distribution handbook for scientists and engineers*, Birkhäuser.

Johnson, N.L., Kemp, A.W. & Kotz, S., 2005. *Univariate Discrete Distributions* 3rd ed., Wiley-Interscience.

Relationship to Other Distributions

The Binomial distribution counts the number of successes in n independent observations of a Bernoulli process.

Binomial Distribution

$Binom(k'|n, p)$

Let

$$K_i \sim Bernoulli(k_i; p) \qquad and \qquad Y = \sum_{i=1}^{n} K_i$$

Then

$$Y \sim Binom(k' = \sum k_i | n, p) \quad \text{where } k' \in \{1, 2, \dots, n\}$$

Special Case:

$$Bernoulli(k; p) = Binom(k; p | n = 1)$$

5.2. Binomial Discrete Distribution

Probability Density Function - f(k)

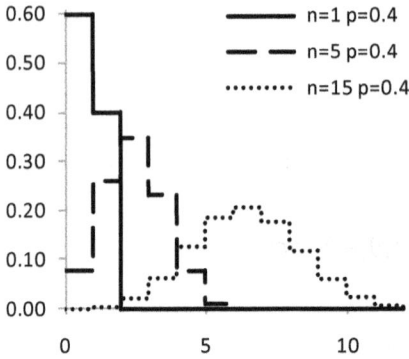

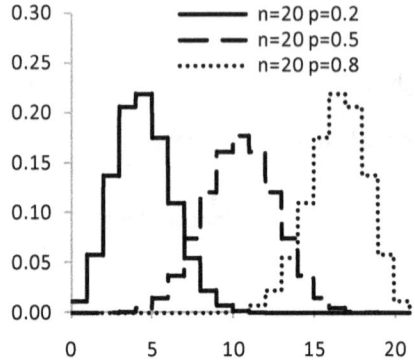

Cumulative Density Function - F(k)

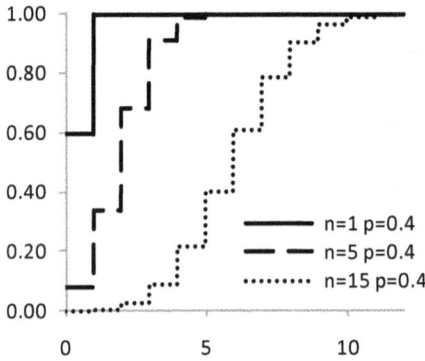

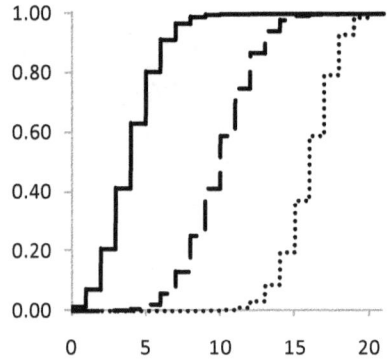

Hazard Rate - h(k)

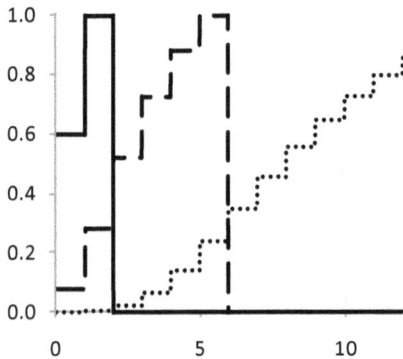

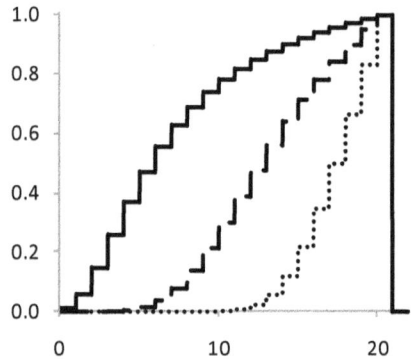

Parameters & Description		
Parameters	n $n \in \{1, 2 \ldots, \infty\}$	Number of Trials.
	p $0 \leq p \leq 1$	Bernoulli probability parameter. Probability of success in a single trial.
Random Variable	$k \in \{0, 1, 2 \ldots, n\}$	

Calculates the probability of getting exactly k successes in n trials.

Distribution	Formulas

PDF

$$f(k) = \binom{n}{k} p^k (1-p)^{n-k}$$

where k combinations from n:

$$\binom{n}{k} = C_k^n = \frac{n!}{k!\,(n-k)!} = \frac{n}{k} C_{k-1}^{n-1}$$

CDF

$$F(k) = \sum_{j=0}^{k} \frac{n!}{j!\,(n-j)!} p^j (1-p)^{n-j}$$
$$= I_{1-p}(n-k, k+1)$$

where $I_p(a,b)$ is the Regularized Incomplete Beta function. See section 1.6.3.

When $n \geq 20$ and $p \leq 0.05$, or if $n \geq 100$ and $np \leq 10$, this can be approximated by a Poisson distribution with $\mu = np$:

$$F(k) \cong e^{-\mu} \sum_{j=0}^{k} \frac{\mu^j}{j!} = \frac{\Gamma(k+1, \mu)}{k!} \cong F_{\chi^2}(2\mu, 2k+2)$$

When $np \geq 10$ and $np(1-p) \geq 10$ then the cdf can be approximated using a normal distribution:

$$F(k) \cong \Phi\left(\frac{k + 0.5 - np}{\sqrt{np(1-p)}}\right)$$

Reliability

$$R(k) = 1 - \sum_{j=0}^{k} \frac{n!}{j!\,(n-j)!} p^j (1-p)^{n-j}$$
$$= \sum_{j=k+1}^{n} \frac{n!}{j!\,(n-j)!} p^j (1-p)^{n-j}$$
$$= I_p(k+1, n-k)$$

where $I_p(a,b)$ is the Regularized Incomplete Beta function. See Section 1.6.3.

Hazard Rate

$$h(k) = \left[1 + \frac{(1+\theta)^n - \sum_{j=0}^{k} \binom{n}{k} \theta^j}{\binom{n}{k} \theta^k}\right]^{-1}$$

where, $\theta = \frac{p}{1-p}$ (Gupta et al. 1997)

Properties and Moments	
Median	$k_{0.5}$ is either $\{\lfloor np \rfloor, \lceil np \rceil\}$
Mode	$\lfloor (n+1)p \rfloor$
Mean - 1st Raw Moment	np
Variance - 2nd Central Moment	$np(1-p)$
Skewness - 3rd Central Moment	$\dfrac{1-2p}{\sqrt{np(1-p)}}$
Excess kurtosis - 4th Central Moment	$\dfrac{6p^2 - 6p + 1}{np(1-p)}$
Characteristic Function	$\left(1 - p + pe^{it}\right)^n$
100α% Percentile Function	Numerically solve for k (which is not arduous for $n \leq 10$): $$k_\alpha = F^{-1}(n,p)$$ For $np \geq 10$ and $np(1-p) \geq 10$ the normal approximation may be used: $$k_\alpha \cong \left\lfloor \Phi^{-1}(\alpha)\sqrt{np(1-p)} + np - 0.5 \right\rfloor$$

Parameter Estimation

Maximum Likelihood Function

Likelihood Function

For complete data only:

$$L(p|E) = \prod_{i=1}^{n_B} \binom{n_i}{k_i} p^{k_i}(1-p)^{n_i - k_i}$$
$$= p^{\Sigma k_i}(1-p)^{\Sigma n_i - \Sigma k_i}$$

Where n_B is the number of Binomial processes, $\Sigma k_i = \sum_{i=1}^{n_B} k_i$, $\Sigma n_i = \sum_{i=1}^{n_B} n_i$ and the combinatory term is ignored (see Section 1.1.6 for discussion).

$\dfrac{dL}{dp} = 0$

solve for p

$$\frac{dL}{dp} = \Sigma k_i \cdot p^{\Sigma(k_i)-1}(1-p)^{\Sigma n_i - \Sigma k_i} - (\Sigma n_i - \Sigma k_i)p^{\Sigma k_i}(1-p)^{\Sigma n_i - 1 - \Sigma k_i}$$

$$\Sigma k_i \cdot p^{\Sigma(k_i)-1}(1-p)^{\Sigma n_i - \Sigma k_i} = (\Sigma n_i - \Sigma k_i)p^{\Sigma k_i}(1-p)^{-1+\Sigma n_i - \Sigma k_i}$$

$$\Sigma k_i \cdot p^{-1} = (\Sigma n_i - \Sigma k_i)(1-p)^{-1}$$

$$\frac{(1-p)}{p} = \frac{\Sigma n_i - \Sigma k_i}{\Sigma k_i}$$

$$p = \frac{\Sigma k_i}{\Sigma n_i}$$

MLE Point Estimates

The MLE point estimate for p:

$$\hat{p} = \frac{\Sigma k_i}{\Sigma n_i}$$

Fisher Information

$$I(p) = \frac{1}{p(1-p)}$$

Confidence Intervals

The confidence intervals for the binomial distribution parameter p is a controversial subject which is still debated. The Wilson interval is recommended for small and large n. (Brown et al. 2001)

$$\overline{p} = \frac{n\hat{p} + \kappa^2/2}{n + \kappa^2} + \frac{\kappa\sqrt{\kappa^2 + 4n\hat{p}(1-\hat{p})}}{2(n+\kappa^2)}$$

$$\underline{p} = \frac{n\hat{p} + \kappa^2/2}{n + \kappa^2} - \frac{\kappa\sqrt{\kappa^2 + 4n\hat{p}(1-\hat{p})}}{2(n+\kappa^2)}$$

where

$$\kappa = \Phi^{-1}\left(\frac{\gamma + 1}{2}\right)$$

It should be noted that most textbooks use the Wald interval (normal approximation) given below, however many articles have shown these estimates to be erratic and cannot be trusted. (Brown et al. 2001)

$$\overline{p} = \hat{p} + \kappa\sqrt{\frac{\hat{p}(1-\hat{p})}{n}}$$

$$\underline{p} = \hat{p} - \kappa\sqrt{\frac{\hat{p}(1-\hat{p})}{n}}$$

For a comparison of binomial confidence interval estimates the reader is referred to (Brown et al. 2001). The following webpage has links to online calculators which use many different methods.

http://en.wikipedia.org/wiki/Binomial_proportion_confidence_interval

Bayesian

Non-informative Priors for p given n, $\pi(p|n)$
(Yang and Berger 1998, p.6)

Type	Prior	Posterior
Uniform Proper Prior with limits $p \in [a,b]$	$\dfrac{1}{b-a}$	Truncated Beta Distribution For a $\le p \le$ b $c.Beta(p; 1+k, 1+n-k)$ Otherwise $\pi(p) = 0$
Uniform Improper Proir with limits $p \in [0,1]$	$1 = Beta(p; 1,1)$	$Beta(p; 1+k, 1+n-k)$
Jeffrey's Prior Reference Prior	$\dfrac{1}{\sqrt{p(1-p)}} = Beta\left(p; \dfrac{1}{2}, \dfrac{1}{2}\right)$	$Beta\left(p; \dfrac{1}{2}+k, \dfrac{1}{2}+n-k\right)$ when $p \in [0,1]$
MDIP	$1.6186p^p(1-p)^{1-p}$	Proper - No Closed Form
Novick and Hall	$p^{-1}(1-p)^{-1} = Beta(0,0)$	$Beta(p; k, n-k)$ when $p \in [0,1]$

Conjugate Priors

UOI	Likelihood Model	Evidence	Dist. of UOI	Prior Para	Posterior Parameters
p from $Binom(k; p, n)$	Binomial	k failures in n trial	Beta	α_o, β_o	$\alpha = \alpha_o + k$ $\beta = \beta_o + n - k$

Description, Limitations and Uses

Example | Five machines are measured for performance on demand. The machines can either fail or succeed in their application. The machines are tested for 10 demands with the following data for each machine:

Machine/Trail	1	2	3	4	5	6	7	8	9	10
1		F = 3					S = 7			
2		F=2					S=8			
3		F=2					S=8			
4		F=3					S=7			
5		F=2					S=8			
μ_i		$n\hat{p}$					$n(1-\hat{p})$			

Assuming machines are homogeneous estimate the parameter p:

Using MLE:

$$\hat{p} = \frac{\sum k_i}{\sum n_i} = \frac{12}{50} = 0.24$$

90% confidence intervals for p:
$$\kappa = \Phi^{-1}(0.95) = 1.64485$$

$$p_{lower} = \frac{n\hat{p} + \kappa^2/2}{n + \kappa^2} - \frac{\kappa\sqrt{\kappa^2 + 4n\hat{p}(1-\hat{p})}}{2(n + \kappa^2)} = 0.1557$$

$$p_{upper} = \frac{n\hat{p} + \kappa^2/2}{n + \kappa^2} + \frac{\kappa\sqrt{\kappa^2 + 4n\hat{p}(1-\hat{p})}}{2(n + \kappa^2)} = 0.351$$

A Bayesian point estimate using a uniform prior distribution $Beta(1, 1)$, with posterior $Beta(p; 13, 39)$ has a point estimate:

$$\hat{p} = \mathrm{E}[Beta(p; 13, 39)] = \frac{13}{52} = 0.25$$

With 90% confidence interval using inverse Beta cdf:

$$[F_{Beta}^{-1}(0.05) = 0.1579, \qquad F_{Beta}^{-1}(0.95) = 0.3532]$$

The probability of observing no failures in the next 10 trials with replacement is:
$$f(0; 10, 0.25) = 0.0563$$

The probability of observing less than 5 failures in the next 10 trials with replacement is:
$$f(0; 10, 0.25) = 0.9803$$

Characteristics

CDF Approximations. The Binomial distribution is one of the most widely used distributions throughout history. Although simple, the CDF function was tedious to calculate prior to the use of computers. As a result, approximations using the Poisson and Normal distribution have been used. For details see 'Related Distributions'.

With Replacement. The Binomial distribution is used to model probability of k successes in n Bernoulli trials. However, the k successes can occur anywhere among the n trials with $_nC_k$ different combinations. Therefore, the Binomial distribution assumes replacement. The equivalent distribution which assumes without replacement is the hypergeometric distribution.

Symmetrical. The distribution is symmetrical when $p = 0.5$.

Compliment. $f(k; n, p) = f(n - k; n, 1 - p)$. Tables usually only provide values up to $n/2$ allowing the reader to calculate to n using the compliment formula.

Assumptions. The binomial distribution describes the behavior of a count variable K if the following conditions apply:
1. The number of observations n is fixed.
2. Each observation is independent.

3. Each observation represents one of two outcomes ("success" or "failure").
4. The probability of "success" is the same for each outcome.

$$K \sim Binom(n, p)$$

Convolution Property

$$\sum K_i \sim Binom(\sum n_i, p)$$

When p is fixed.

Applications   Used to model independent repeated trials which have two outcomes. Examples used in Reliability Engineering are:
- Number of independent components failed, k, from a population, n after receiving a shock.
- Number of failures to start, k, from n demands on a component.
- Number of independent items found defective, k, from a population of n items.

Resources   Online:
http://mathworld.wolfram.com/BinomialDistribution.html
http://en.wikipedia.org/wiki/Binomial_distribution
http://socr.ucla.edu/htmls/SOCR_Distributions.html (web calc)

Books:
Collani, E.V. & Dräger, K., 2001. *Binomial distribution handbook for scientists and engineers*, Birkhäuser.

Johnson, N.L., Kemp, A.W. & Kotz, S., 2005. *Univariate Discrete Distributions* 3rd ed., Wiley-Interscience.

Relationship to Other Distributions

The Binomial distribution counts the number of successes k in n independent observations of a Bernoulli process.

Let

Bernoulli Distribution $$K_i \sim Bernoulli(k'_i; p) \quad and \quad Y = \sum_{i=1}^{n} K_i$$

Bernoulli$(k'; p)$ Then

$$Y \sim Binom(\sum k'_i; n, p) \quad where \ k \in \{1, 2, \dots, n\}$$

Special Case:
$$Bernoulli(k; p) = Binom(k; p | n = 1)$$

Hypergeometric Distribution $HyperGeom$ $(k; n, m, N)$	The hypergeometric distribution is used to model probability of k successes in n Bernoulli trials from a population N, with m successors without replacement. $$f(k; n, m, N)$$ Limiting Case for $n \gg k$ and p not near 0 or 1: $$\lim_{n \to \infty} Binom(k; n, p = \frac{m}{N}) = HyperGeom(k; n, m, N)$$

Limiting Case for constant p:
$$\lim_{\substack{n \to \infty \\ p=p}} Binom(k|n, p) = Norm\big(k|\mu = np, \sigma^2 = np(1 - p)\big)$$

Normal Distribution

$Norm(t; \mu, \sigma^2)$

The Normal distribution can be used as an approximation of the Binomial distribution when $np \geq 10$ and $np(1 - p) \geq 10$.

$$Binom(k|p, n) \approx Norm\big(k + 0.5|\mu = np, \sigma^2 = np(1 - p)\big)$$

Limiting Case for constant np:
$$\lim_{\substack{n \to \infty \\ np=\mu}} Binom(k; n, p) = Pois(k; \mu = np)$$

The Poisson distribution is the limiting case of the Binomial distribution when n is large but the ratio of np remains constant. Hence the Poisson distribution can model rare events.

The Poisson distribution can be used as an approximation to the Binomial distribution when $n \geq 20$ and $p \leq 0.05$, or if $n \geq 100$ and $np \leq 10$.

Poisson Distribution

$Pois(k; \mu)$

The Binomial is expressed in terms of the total number of a probability of success, p, and trials, N. Where a Poisson distribution is expressed in terms of a success rate and does not need to know the total number of trials.

The derivation of the Poisson distribution from the binomial can be found at http://mathworld.wolfram.com/PoissonDistribution.html.

This interpretation can also be used to understand the conditional distribution of a Poisson random variable:
Let
$$K_1, K_2 \sim Pois(\mu)$$
Given
$$n = K_1 + K_2 = number\ of\ events$$
Then
$$K_1|n \sim Binom\left(k; n, p = \frac{\mu_1}{\mu_1 + \mu_2}\right)$$

Multinomial Distribution

$MNom_d(\mathbf{k}|n, \mathbf{p})$

Special Case:
$$MNom_{d=2}(\mathbf{k}|n, \mathbf{p}) = Binom(k|n, p)$$

5.3. Poisson Discrete Distribution

Probability Density Function - f(k)

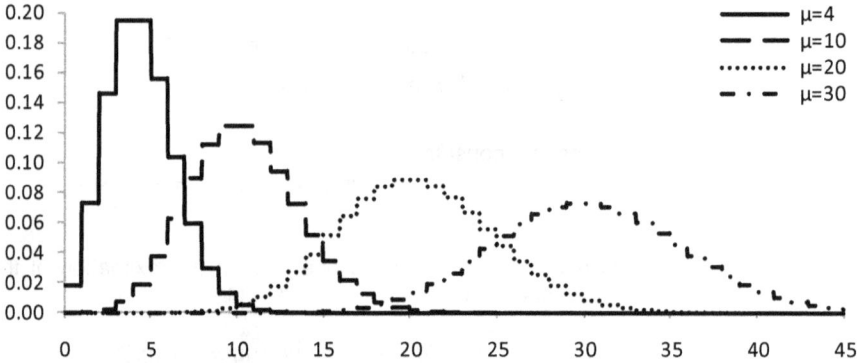

Cumulative Density Function - F(k)

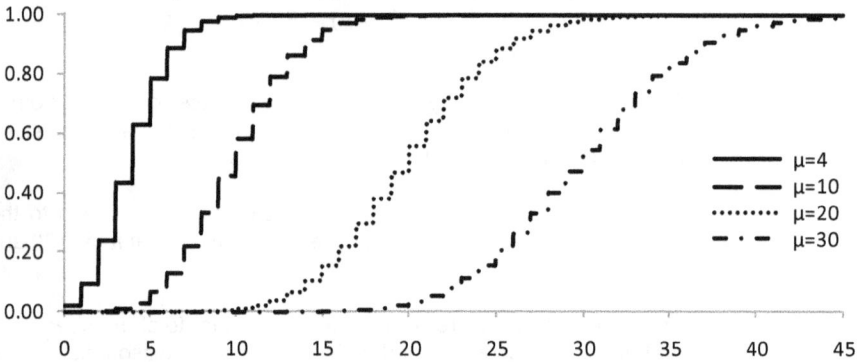

Hazard Rate - h(k)

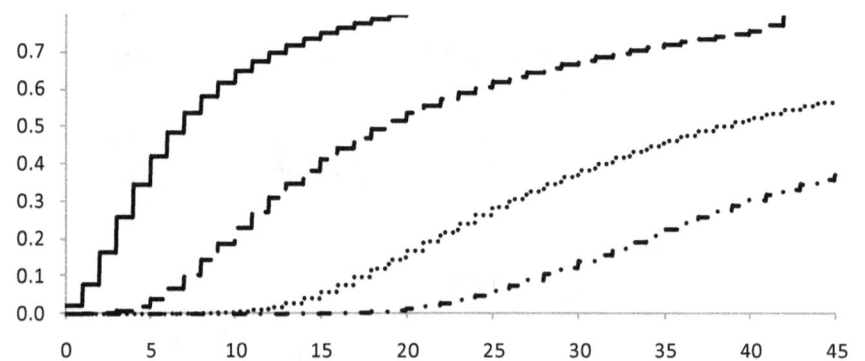

Parameters & Description

Parameters	μ	$\mu > 0$	Shape Parameter: The value of μ is the expected number of events per time, over space or other physical dimensions. If the Poisson distribution is modeling failure events over time, then $\mu = \lambda t$ is the average number of failures that would occur in time t. In this case t is fixed and λ becomes the distribution parameter. Parameter λ is referred to a "rate" when events occur over time, and is referred to as "intensity" when events occur over space. Some texts use the symbol ρ instead.
Random Variable		k is an integer, $k \geq 0$	

Distribution — Formulas

PDF

$$f(k) = \frac{\mu^k}{k!} e^{-\mu} = \frac{(\lambda t)^k}{k!} e^{-\lambda t}$$

CDF

$$F(k) = e^{-\mu} \sum_{j=0}^{k} \frac{\mu^j}{j!} = \frac{\Gamma(k+1, \mu)}{k!}$$
$$= F_{\chi^2}(2\mu, 2k+2)$$

Where $F_{\chi^2}(x|v)$ is the Chi-square CDF.

When $\mu > 10$ the $F(k)$ can be approximated by a normal distribution:
$$F(k) \cong \Phi\left(\frac{k + 0.5 - \mu}{\sqrt{\mu}}\right)$$

Reliability

$$R(k) = 1 - F(k)$$

Hazard Rate

$$h(k) = \left[1 + \frac{k!}{\mu}\left(e^\mu - 1 - \sum_{j=1}^{k} \frac{\mu^j}{j!}\right)\right]^{-1}$$

(Gupta et al. 1997)

Properties and Moments

Median

See 100α% Percentile Function when $\alpha = 0.5$.

Mode

$\lfloor \mu \rfloor$
where $\lfloor \mu \rfloor$ is the floor function[3]

---

[3] $\lfloor \mu \rfloor$ = is the floor function (largest integer not greater than μ)

Mean - 1st Raw Moment	μ
Variance - 2nd Central Moment	μ
Skewness - 3rd Central Moment	$1/\sqrt{\mu}$
Excess kurtosis - 4th Central Moment	$1/\mu$
Characteristic Function	$\exp\{\mu(e^{ik}-1)\}$

100α% Percentile Function

Numerically solve for k (which is not arduous for $\mu \leq 10$):
$$k_\alpha = F^{-1}(\alpha)$$
For $k > 10$ the normal approximation may be used:
$$k_\alpha \cong \left\lfloor \sqrt{\mu}\Phi^{-1}(\alpha) + \mu - 0.5 \right\rfloor$$

Parameter Estimation

Maximum Likelihood Estimates

Likelihood Functions

For complete data:
$$L(\mu|E) = \underbrace{\prod_{i=1}^{n} \frac{\mu^{k_i^F}}{k_i^F!} e^{-\mu}}_{\text{known k}}$$

where n is the total number of events and k is the number events of interest.

Log-Likelihood Function

$$\Lambda = \underbrace{-n\mu + \sum_{i=1}^{n} \{k_i \ln(\mu) - \ln(k_i!)\}}_{\text{known k}}$$

$\dfrac{\partial \Lambda}{\partial \mu} = 0$

$$\underbrace{\frac{\partial \Lambda}{\partial \mu} = -n + \frac{1}{\mu}\sum_{i=1}^{n} k_i = 0}_{\text{known k}}$$

MLE Point Estimates

For complete data solving $\dfrac{\partial \Lambda}{\partial \mu} = 0$ gives:
$$\hat{\mu} = \frac{1}{n}.\sum_{i=1}^{n} k_i \quad or \quad \hat{\lambda} = \frac{1}{tn}.\sum_{i=1}^{n} k_i$$

Note that in this context:
t = the unit of time for which the rate, λ is being measured.
n = the number of Poisson processes (number of events) for which the exact number of events of interest (e.g., failures), k, was known.
k_i = the number of failures that occurred within the ith event.

When there is only one Poisson process this reduces to:
$$\hat{\mu} = k \quad or \quad \hat{\lambda} = \frac{k}{t}$$
For censored data numerical methods are needed to maximize the log-likelihood function.

Fisher Information		$I(\lambda) = \dfrac{1}{\lambda}$	

		λ_{lower} - 2 Sided	λ_{upper} - 2 Sided
100γ% Confidence Interval	Conservative two-sided confidence intervals.	$\dfrac{\chi^2_{\left[\frac{1-\gamma}{2}\right]}(2\sum k_i)}{2tn}$	$\dfrac{\chi^2_{\left[\frac{1+\gamma}{2}\right]}(2\sum k_i + 2)}{2tn}$
(complete data only)	When k is large $(k > 10)$ two sided intervals	$\hat{\lambda} - \Phi^{-1}\left(\dfrac{1+\gamma}{2}\right)\sqrt{\dfrac{\hat{\lambda}}{tn}}$	$\hat{\lambda} + \Phi^{-1}\left(\dfrac{1+\gamma}{2}\right)\sqrt{\dfrac{\hat{\lambda}}{tn}}$

(Nelson 1982, p.201) Note: The first confidence intervals are conservative in that at least 100γ%. Exact confidence intervals cannot be easily achieved for discrete distributions.

Bayesian

Non-informative Priors $\pi(\lambda)$ in known time interval t

Type	Prior	Posterior
Uniform Proper Prior with limits $\lambda \in [a,b]$	$\dfrac{1}{b-a}$	Truncated Gamma Distribution For a $\leq \lambda \leq$ b $\quad c.\,Gamma(\lambda; 1 + k, t)$ Otherwise $\pi(\lambda) = 0$
Uniform Improper Prior with limits $\lambda \in [0, \infty)$	$1 \propto Gamma(1,0)$	$Gamma(\lambda; 1 + k, t)$
Jeffrey's Prior	$\dfrac{1}{\sqrt{\lambda}} \propto Gamma(\tfrac{1}{2}, 0)$	$Gamma(\lambda; \tfrac{1}{2} + k, t)$ when $\lambda \in [0, \infty)$
Novick and Hall	$\dfrac{1}{\lambda} \propto Gamma(0,0)$	$Gamma(\lambda; k, t)$ when $\lambda \in [0, \infty)$

Conjugate Priors

UOI	Likelihood Model	Evidence	Dist. of UOI	Prior Para	Posterior Parameters
λ from $Pois(k; \mu)$	Exponential	n_F failures in t_T unit of time	Gamma	k_0, Λ_0	$k = k_o + n_F$ $\Lambda = \Lambda_o + t_T$

Description, Limitations and Uses

Example	Three vehicle tires were run on a test area for 1000km have punctures at the following distances: Tire 1: No punctures Tire 2: 400km, 900km Tire 3: 200km

Punctures can be modeled as a renewal process with perfect repair and an inter-arrival time modeled by an exponential distribution. Due to the Poisson distribution being homogeneous in time, the test from multiple tires can be combined and considered a test of one tire with multiple renewals. See example in section 1.1.6.

Total time on test is $3 \times 1000 = 3000$km. Total number of failures is 3. Therefore, using MLE the estimate of λ:

$$\hat{\lambda} = \frac{k}{t_T} = \frac{3}{3000} = 1E\text{-}3$$

With 90% confidence interval (conservative):
$$\left[\frac{\chi^2_{(0.05)}(6)}{6000} = 0.272E\text{-}3, \qquad \frac{\chi^2_{(0.95)}(8)}{6000} = 2.584E\text{-}3 \right]$$

A Bayesian point estimate using the Jeffery non-informative improper prior $Gamma(\frac{1}{2}, 0)$, with posterior $Gamma(\lambda; 3.5, 3000)$ has a point estimate:
$$\hat{\lambda} = E[Gamma(\lambda; 3.5, 3000)] = \frac{3.5}{3000} = 1.16E - 3$$

With 90% confidence interval using inverse Gamma cdf:
$$[F_G^{-1}(0.05) = 0.361E\text{-}3, \qquad F_G^{-1}(0.95) = 2.344E\text{-}3]$$

Characteristics

The Poisson distribution is also known as the Rare Event distribution.

If the following assumptions are met than the process follows a Poisson distribution:
- The chance of two simultaneous events is negligible or impossible (such as renewal of a single component);
- The expected value of the random number of events in a region is proportional to the size of the region.
- The random number of events in non-overlapping regions are independent.

μ **characteristics:**
- μ is the expected number of events for the unit of time being measured.
- When the unit of time varies μ can be transformed into a rate and time measure, λt.
- For $\mu \lesssim 10$ the distribution is skewed to the right.
- For $\mu \gtrsim 10$ the distribution approaches a normal distribution with a $\mu = \mu$ and $\sigma = \sqrt{\mu}$.

$$K \sim Pois(\mu)$$
Convolution property
$$K_1 + K_2 + \ldots + K_n \sim Pois(k; \textstyle\sum \mu_i)$$

Applications	**Homogeneous Poisson Process (HPP).** The Poisson distribution gives the distribution of exactly k failures occurring in a HPP. See relation to exponential and gamma distributions.

Renewal Theory. Used in renewal theory as the counting function and may model non-homogeneous (aging) components by using a time dependent failure rate, $\lambda\,(t)$.

Binomial Approximation. Used to model the Binomial distribution when the number of trials is large and μ remains moderate. This can greatly simplify Binomial distribution calculations.

Rare Event. Used to model rare events when the number of trials is large compared to the rate at which events occur.

Resources

Online:
http://mathworld.wolfram.com/PoissonDistribution.html
http://en.wikipedia.org/wiki/Poisson_distribution
http://socr.ucla.edu/htmls/SOCR_Distributions.html (interactive web calculator)

Books:
Haight, F.A., 1967. Handbook of the Poisson distribution [by] Frank A. Haight, New York, Wiley.

Nelson, W.B., 1982. Applied Life Data Analysis, Wiley-Interscience.

Johnson, N.L., Kemp, A.W. & Kotz, S., 2005. Univariate Discrete Distributions 3rd ed., Wiley-Interscience.

Relationship to Other Distributions

Let
$$K \sim Pois(\text{k}; \mu = \lambda t)$$
Given
$$time = T_1 + T_2 + \cdots + T_K + T_{K+1} \cdots$$

Exponential Distribution

Then
$$T_1, T_2 \ldots \sim Exp(\text{t}; \lambda)$$

$Exp(t; \lambda)$

The time between each arrival of T is exponentially distributed.

Special Cases:
$$Pois(\text{k}; \lambda t | k = 1) = Exp(t; \lambda)$$

Let
$$T_1 \ldots T_k \sim Exp(\lambda) \qquad and \qquad T_t = T_1 + T_2 + \cdots + T_k$$

Gamma Distribution

Then
$$T_t \sim Gamma(k, \lambda)$$

$Gamma(k|\lambda)$

The Poisson distribution is the probability that exactly k failures have been observed in time t. This is the probability that t is between T_k and T_{k+1}.

$$f_{Poisson}(k; \lambda t) = \int_{k}^{k+1} f_{Gamma}(t; x, \lambda)dx$$
$$= F_{Gamma}(t; k+1, \lambda) - F_{Gamma}(t; k, \lambda)$$

where k is an integer.

Limiting Case for constant np:

$$\lim_{\substack{n \to \infty \\ np=\mu}} Binom(k; n, p) = Pois(k|\mu = np)$$

The Poisson distribution is the limiting case of the Binomial distribution when n is large but the ratio of np remains constant. Hence, the Poisson distribution can model rare events.

The Poisson distribution can be used as an approximation to the Binomial distribution when $n \geq 20$ and $p \leq 0.05$, or if $n \geq 100$ and $np \leq 10$.

Binomial Distribution

$Binom(k|p, N)$

The Binomial is expressed in terms of the total number of a probability of success, p, and trials, N. Where a Poisson distribution is expressed in terms of a success rate and does not need to know the total number of trials.

The derivation of the Poisson distribution from the binomial can be found at *http://mathworld.wolfram.com/PoissonDistribution.html*.

This interpretation can also be used to understand the conditional distribution of a Poisson random variable:
Let

$$K_1, K_2 \sim Pois(\mu)$$

Given

$$n = K_1 + K_2 = number\ of\ events$$

Then

$$K_1|n \sim Binom\left(k; n \middle| p = \frac{\mu_1}{\mu_1 + \mu_2}\right)$$

$$\lim_{\mu \to \infty} F_{Poisson}(k; \mu) = F_{Normal}(k; \mu' = \mu, \sigma^2 = \mu)$$

Normal Distribution

$Norm(k|\mu', \sigma)$

This is a good approximation when $\mu > 1000$. When $\mu > 10$ the same approximation can be made with a correction:

$$\lim_{\mu \to \infty} F_{Poisson}(k; \mu) = F_{Normal}(k; \mu' = \mu - 0.5, \sigma^2 = \mu)$$

Chi-square Distribution

$\chi^2(t|v)$

$$Pois(k|\mu) = \chi^2(x = 2\mu, v = 2k + 2)$$

6. Bivariate and Multivariate Distributions

6.1. Bivariate Normal Continuous Distribution

Probability Density Function - f(x,y)

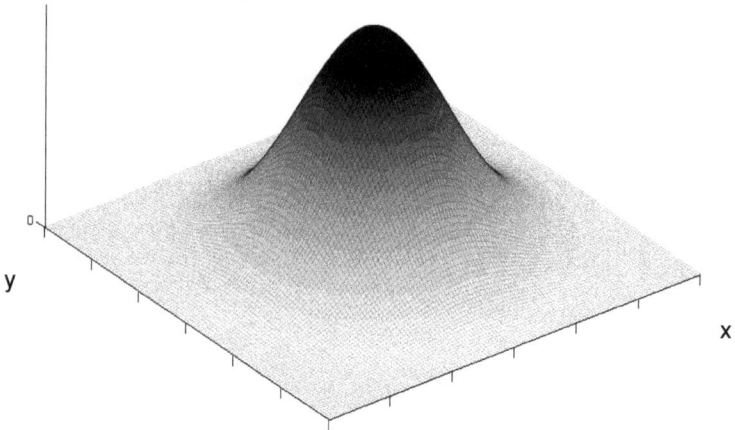

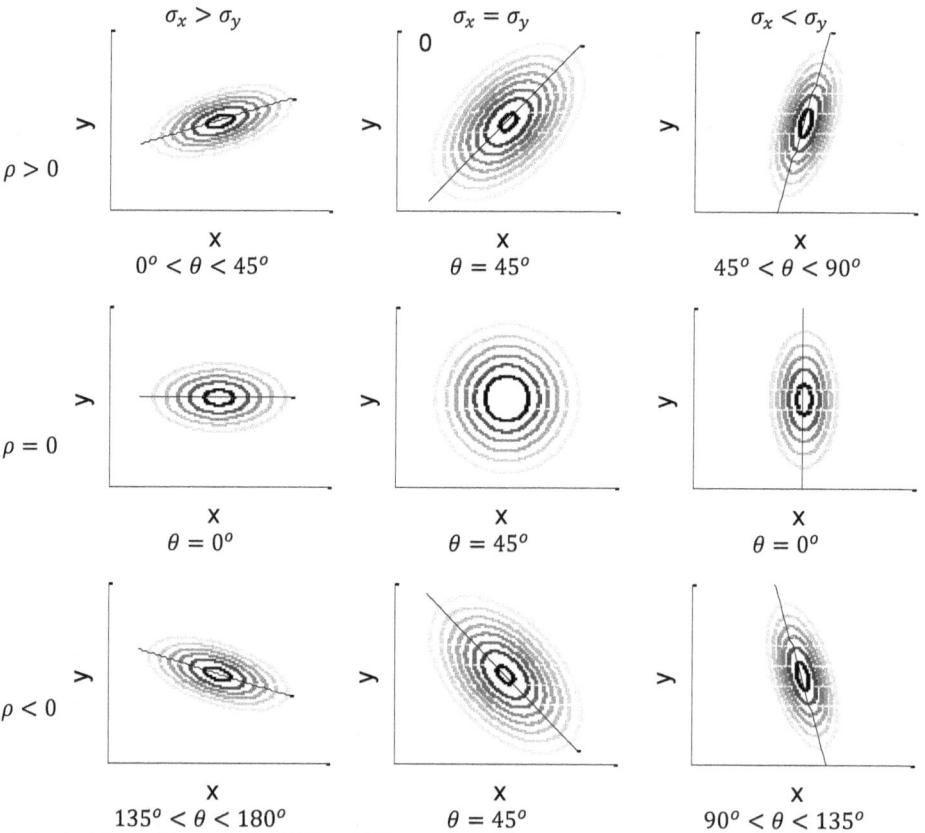

Adapted from (Kotz et al. 2000, p.256)

	Parameters & Description	
Parameters	μ_x, μ_y $\begin{array}{c}-\infty < \mu_j < \infty \\ j \in \{x, y\}\end{array}$	*Location parameter:* The mean of each random variable.
	σ_x, σ_y $\begin{array}{c}\sigma_j > 0 \\ j \in \{x, y\}\end{array}$	*Scale parameter:* The standard deviation of each random variable.
	ρ $-1 \le \rho \le 1$	*Correlation Coefficient:* The correlation between the two random variables. $\rho = corr(X, Y) = \dfrac{cov[XY]}{\sigma_x \sigma_y}$ $= \dfrac{E[(X - \mu_x)(Y - \mu_y)]}{\sigma_x \sigma_y}$

Limits $-\infty < x < \infty \quad and \quad -\infty < y < \infty$

Distribution	**Formulas**

PDF

$$f(x,y) = \frac{1}{2\pi\sigma_x\sigma_y\sqrt{1-\rho^2}} \exp\left[\frac{z_x^2 + z_y^2 - 2\rho z_x z_y}{-2(1-\rho^2)}\right]$$
$$= \phi(x)\phi(y|x)$$
$$= \phi(x)\phi\left(\frac{y - \rho x}{\sqrt{1-\rho^2}}\right) = \phi(y)\phi\left(\frac{x - \rho y}{\sqrt{1-\rho^2}}\right)$$

where, ϕ is the standard normal distribution and:
$$z_j = \frac{x - \mu_j}{\sigma_j} \qquad j \in \{x, y\}$$

Marginal PDF

$$f(x) = \int_{-\infty}^{\infty} f(x, y)\, dy \qquad\qquad f(y) = \int_{-\infty}^{\infty} f(x, y)\, dx$$
$$= \frac{1}{\sigma_x\sqrt{2\pi}} \exp\left[-\frac{1}{2}(z_x)^2\right] \qquad = \frac{1}{\sigma_y\sqrt{2\pi}} \exp\left[-\frac{1}{2}(z_y)^2\right]$$
$$= Norm(\mu_x, \sigma_x) \qquad\qquad\qquad = Norm(\mu_y, \sigma_y)$$

Conditional PDF

$$f(x|y) = Norm\left(\mu_{x|y} = \mu_x + \rho\left(\frac{\sigma_x}{\sigma_y}\right)(y - \mu_y),\ \sigma_{x|y}^2 = \sigma_x^2(1-\rho^2)\right)$$
$$f(y|x) = Norm\left(\mu_{y|x} = \mu_y + \rho\left(\frac{\sigma_y}{\sigma_x}\right)(y - \mu_x),\ \sigma_{y|x}^2 = \sigma_y^2(1-\rho^2)\right)$$

CDF

$$F(x,y) = \frac{1}{2\pi\sigma_x\sigma_y\sqrt{1-\rho^2}} \int_{-\infty}^{x}\int_{-\infty}^{y} \exp\left[\frac{z_u^2 + z_v^2 - 2\rho z_u z_v}{-2(1-\rho^2)}\right] du\, dv$$

where
$$z_j = \frac{x - \mu_j}{\sigma_j}$$

Reliability	$$R(x,y) = \frac{1}{2\pi\sigma_x\sigma_y\sqrt{1-\rho^2}}\int_x^\infty\int_y^\infty \exp\left[\frac{z_u^2 + z_v^2 - 2\rho z_u z_v}{2(1-\rho^2)}\right] du\, dv$$
	where

$$z_j = \frac{x - \mu_j}{\sigma_j}$$

Properties and Moments

Median	$\begin{bmatrix}\mu_x \\ \mu_y\end{bmatrix}$
Mode	$\begin{bmatrix}\mu_x \\ \mu_y\end{bmatrix}$
Mean - 1st Raw Moment	$E\begin{bmatrix}X \\ Y\end{bmatrix} = \begin{bmatrix}\mu_x \\ \mu_y\end{bmatrix}$

The mean of the marginal distributions is:
$$E[X] = \mu_x$$
$$E[Y] = \mu_y$$

The mean of the conditional distributions gives the following lines (also called the regression lines):
$$E(X|Y = y) = \mu_x + \rho.\frac{\sigma_x}{\sigma_y}(y - \mu_y)$$
$$E(Y|X = x) = \mu_y + \rho.\frac{\sigma_y}{\sigma_x}(y - \mu_x)$$

Variance - 2nd Central Moment	$Cov\begin{bmatrix}X \\ Y\end{bmatrix} = \begin{bmatrix}\sigma_1^2 & \rho\sigma_1\sigma_2 \\ \rho\sigma_1\sigma_2 & \sigma_2^2\end{bmatrix}$

Variance of marginal distributions:
$$Var(X) = \sigma_x^2$$
$$Var(Y) = \sigma_y^2$$

Variance of conditional distributions:
$$Var(X|Y = y) = \sigma_x^2(1 - \rho^2)$$
$$Var(Y|X = x) = \sigma_y^2(1 - \rho^2)$$

100α% Percentile Function	An ellipse containing 100α % of the distribution is (Kotz et al. 2000, p.254):

$$\frac{(z_x^2 + z_y^2 - 2\rho z_x z_y)}{-2(1-\rho^2)} = \ln(1 - \alpha)$$

where
$$z_j = \frac{x - \mu_j}{\sigma_j} \qquad j \in \{x, y\}$$

For the standard bivariate normal:
$$\frac{x^2 + y^2 - 2\rho xy}{-2(1-\rho^2)} = \ln(1 - \alpha)$$

Parameter Estimation

Maximum Likelihood Function

MLE Point
Estimates

When there is only complete failure data the MLE estimates can be given as (Kotz et al. 2000, p.294):

$$\widehat{\mu_x} = \frac{1}{n_F}\sum_{i=1}^{n_F} x_i \qquad \widehat{\sigma_x^2} = \frac{1}{n_F}\sum_{i=1}^{n_F}(x_i - \widehat{\mu_x})^2$$

$$\widehat{\mu_y} = \frac{1}{n_F}\sum_{i=1}^{n_F} y_i \qquad \widehat{\sigma_y^2} = \frac{1}{n_F}\sum_{i=1}^{n_F}(y_i - \widehat{\mu_y})^2$$

$$\hat{\rho} = \frac{1}{\widehat{\sigma_x}\widehat{\sigma_y}n_F}\sum_{i=1}^{n_F}(x_i - \mu_x)(y_i - \mu_y)$$

If one or more of the variables are known, different estimators are given in (Kotz et al. 2000, pp.294-305).

A correction factor of -1 can be introduced to the $\widehat{\sigma^2}$ to give the unbiased estimators:

$$\widehat{\sigma_x^2} = \frac{1}{n_F - 1}\sum_{i=1}^{n_F}(x_i - \widehat{\mu_x})^2 \qquad \widehat{\sigma_y^2} = \frac{1}{n_F - 1}\sum_{i=1}^{n_F}(y_i - \widehat{\mu_y})^2$$

Bayesian

Non-informative Priors: A complete coverage of numerous reference prior distributions with different parameter ordering is contained in (Berger & Sun 2008).

For a summary of the general Bayesian priors and conjugates see the multivariate normal distribution.

Description, Limitations and Uses

Example

The accuracy of a cutting machine used in manufacturing is desired to be measured. 5 cuts at the required length are made. The lengths and room temperature were measured as:

7.436, 10.270, 10.466, 11.039, 11.854 mm
19.51, 21.23, 21.41, 22.78, 26.78 °C

MLE estimates are:

$$\widehat{\mu_x} = \frac{\sum x_i}{n} = 10.213$$

$$\widehat{\mu_T} = \frac{\sum t_i}{n} = 22.342$$

$$\widehat{\sigma_x^2} = \frac{\sum(x_i - \widehat{\mu_L})^2}{n-1} = 2.7885$$

$$\widehat{\sigma_T^2} = \frac{\sum(t_i - \widehat{\mu_T})^2}{n-1} = 7.5033$$

$$\hat{\rho} = \frac{1}{\widehat{\sigma_x}\widehat{\sigma_T}n_F}\sum_{i=1}^{n_F}(x_i - \mu_x)(t_i - \mu_T) = 0.1454$$

If you know the temperature is 24 °C what is the likely cutting distance distribution?

$$f(x|t = 24) = Norm\left(\mu_{x|t} = \mu_x + \rho\left(\frac{\sigma_x}{\sigma_t}\right)(t - \mu_T), \ \sigma_{x|t}^2 = \sigma_x^2(1 - \rho^2)\right)$$

$$f(x|t = 24) = Norm(10.303, \ 2.730)$$

Characteristic   Also known as Binormal Distribution.

Let U, V and W be three independent normally distributed random variables. Then let:

$$X = U + V$$
$$Y = V + W$$

Then (X, Y) has a bivariate normal distribution. (Balakrishnan & Lai 2009, p.483)

Independence. If X and Y are jointly normal random variables, then they are independent when $\rho = 0$. This gives a contour plot of $f(x, y)$ with concentric circles around the origin. When given a value on the y axis it does not assist in estimating the value on the x axis and therefore are independent. When X and Y are independent, the pdf reduces to:

$$f(x, y) = \frac{1}{2\pi\sigma_x\sigma_y} \exp\left[-\frac{z_x^2 + z_y^2}{2}\right]$$

Correlation Coefficient ρ. (Yang et al. 2004, p.49)
-    $\rho > 0$. When X increases then Y also tends to increase. When $\rho = 1$ X and Y have a perfect positive linear relationship such that $Y = c + mX$ where m is positive.
-    $\rho < 0$. When X increases then Y also tends to decrease. When $\rho = -1$ X and Y have a perfect negative linear relationship such that $Y = c + mX$ where m is negative.
-    $\rho = 0$. Increases or decreases in X have no effect on Y. X and Y are independent.

Ellipse Axis. (Kotz et al. 2000, p.254) The slope of the main axis from the x-axis is given as:

$$\theta = \frac{1}{2}tan^{-1}\left[\frac{2\rho\sigma_x\sigma_y}{\sigma_x^2 - \sigma_y^2}\right]$$

If $\sigma_x = \sigma_y$ for positive ρ the main axis of the ellipse is 45° from the x-axis. For negative ρ the main axis of the ellipse is -45° from the x-axis.

Circular Normal Density Function. (Kotz et al. 2000, p.255) When $\sigma_x = \sigma_y$ and $\rho = 0$ the bivariate distribution is known as a circular normal density function.

Elliptical Normal Distribution (Kotz et al. 2000, p.255). If $\rho = 0$ and $\sigma_x \neq \sigma_y$ then the distribution may be known as an elliptical normal distribution.

Standard Bivariate Normal Distribution. Occurs when $\mu = 0$ and $\sigma = 1$. For positive ρ the main axis of the ellipse is 45° from the x-axis. For negative ρ the main axis of the ellipse is -45° from the x-axis.

$$f(x,y) = \frac{1}{2\pi\sqrt{1 - \rho^2}} \exp\left[-\frac{x^2 + y^2 - 2\rho xy}{2(1 - \rho^2)}\right]$$

Mean / Median / Mode:
As per the univariate distributions the mean, median and mode are equal.

Matrix Form. The bivariate distribution may be written in matrix form as:

$$X = \begin{pmatrix} X_1 \\ X_2 \end{pmatrix} \quad \mu = \begin{pmatrix} \mu_1 \\ \mu_2 \end{pmatrix} \quad \Sigma = \begin{bmatrix} \sigma_1^2 & \rho\sigma_1\sigma_2 \\ \rho\sigma_1\sigma_2 & \sigma_2^2 \end{bmatrix}$$

when $X \sim Norm_2(\mu, \Sigma)$

$$f(x) = \frac{1}{2\pi\sqrt{|\Sigma|}} \exp\left[-\frac{1}{2}(x - \mu)^T \Sigma^{-1}(x - \mu)\right]$$

Where $|\Sigma|$ is the determinant of Σ. This is the general form used in multivariate normal distribution.

The following properties are given in matrix form:

Convolution Property
Let $X \sim Norm(\mu_x, \Sigma_x)$ $Y \sim Norm(\mu_y, \Sigma_y)$
Where $X \perp Y$ (independent)
Then $X + Y \sim Norm(\mu_x + \mu_y, \Sigma_x + \Sigma_y)$

Note if **X** and **Y** are dependent then $X + Y$ may not be even be normally distributed.(Novosyolov 2006)

Scaling Property
Let, $Y = AX + b$, Y is a p x 1 matrix b is a p x 1 matrix

Then, $Y \sim Norm(A\mu + b, A\Sigma A^T)$ A is a x 2 matrix

Marginalize Property:
Let $\begin{bmatrix} X_1 \\ X_2 \end{bmatrix} \sim Norm\left(\begin{bmatrix} \mu_1 \\ \mu_2 \end{bmatrix}, \begin{bmatrix} \sigma_1^2 & \rho\sigma_1\sigma_2 \\ \rho\sigma_1\sigma_2 & \sigma_2^2 \end{bmatrix}\right)$

Then $X_1 \sim Norm(\mu_1, \sigma_1)$

Conditional Property:

Let

$$\begin{bmatrix} X_1 \\ X_2 \end{bmatrix} \sim Norm\left(\begin{bmatrix} \mu_1 \\ \mu_2 \end{bmatrix}, \begin{bmatrix} \sigma_1^2 & \rho\sigma_1\sigma_2 \\ \rho\sigma_1\sigma_2 & \sigma_2^2 \end{bmatrix}\right)$$

Then

$$f(x_1|x_2) = Norm(\mu_{1|2}, \sigma_{1|2})$$

Where,

$$\mu_{1|2} = \mu_1 + \rho\left(\frac{\sigma_1}{\sigma_2}\right)(x_2 - \mu_2)$$
$$\sigma_{1|2} = \sigma_1\sqrt{1 - \rho^2}$$

It should be noted that the standard deviation of the marginal distribution does not depend on the given value.

Applications

The bivariate distribution is used in many applications which are common to the multivariate normal distribution. Please refer to multivariate normal distribution for a more complete coverage.

Graphical Representation of Multivariate Normal. As with all bivariate distributions having only two dependent variables allows it to be easily graphed (in a three-dimensional graph) and visualized. As such the bivariate normal is popular in introducing higher dimensional cases.

Resources

Online:
http://mathworld.wolfram.com/BivariateNormalDistribution.html
http://en.wikipedia.org/wiki/Multivariate_normal_distribution
http://www.aiaccess.net/English/Glossaries/GlosMod/e_gm_binormal_distri.htm (interactive visual representation)

Books:
Balakrishnan, N. & Lai, C., 2009. *Continuous Bivariate Distributions* 2nd ed., Springer.

Yang, K. et al., 2004. *Multivariate Statistical Methods in Quality Management* 1st ed., McGraw-Hill Professional.

Patel, J.K, Read, C.B, 1996. *Handbook of the Normal Distribution*, 2nd Edition, CRC

Tong, Y.L., 1990. *The Multivariate Normal Distribution*, Springer.

6.2. Dirichlet Continuous Distribution

Probability Density Function - f(**x**)

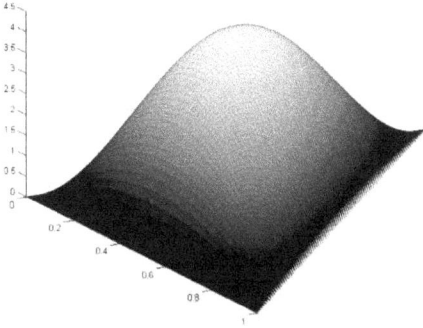

$$Dir_2([x_1, x_2]^T; [2,2,2]^T)$$

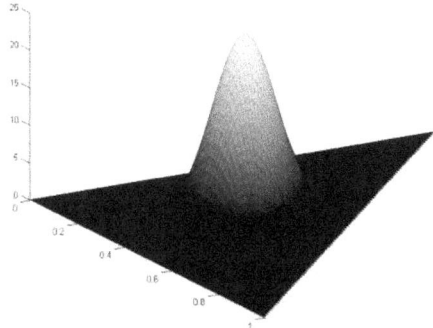

$$Dir_2([x_1, x_2]^T; [10,10,10]^T)$$

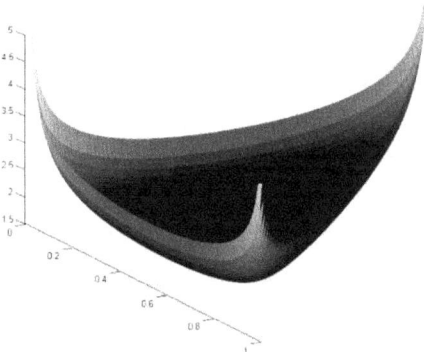

$$Dir_2\left([x_1, x_2]^T; \left[\tfrac{1}{2}, \tfrac{1}{2}, \tfrac{1}{2}\right]^T\right)$$

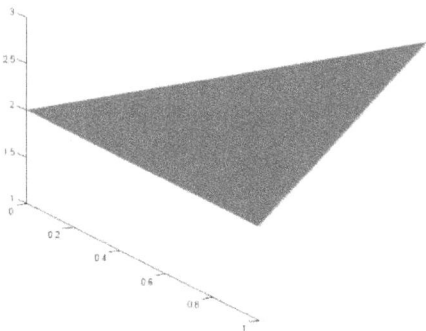

$$Dir_2([x_1, x_2]^T; [1,1,1]^T)$$

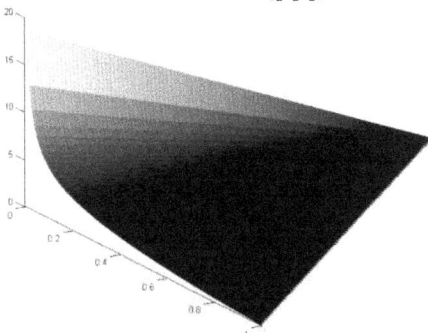

$$Dir_2\left([x_1, x_2]^T; \left[\tfrac{1}{2}, 1, 2\right]^T\right)$$

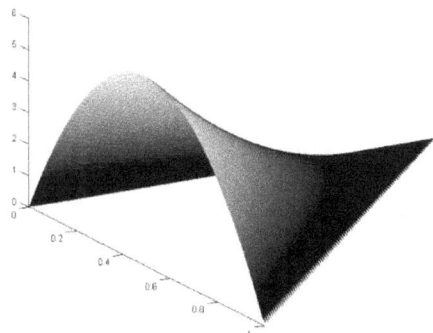

$$Dir_2([x_1, x_2]^T; [2,1,2]^T)$$

Parameters & Description

Parameters			
	$\boldsymbol{\alpha} = [\alpha_1, \alpha_2, ..., \alpha_d, \alpha_0]^T$	$\alpha_i > 0$	*Shape Matrix*. Note that the matrix α is $d + 1$ in length.
	d	$d \geq 1$ (integer)	*Dimension.* The number of random variables being modeled.

Limits	
	$0 \leq x_i \leq 1$
	$\displaystyle\sum_{i=1}^{d} x_i \leq 1$

Distribution	Formulas

PDF

$$f(\mathbf{x}) = \frac{1}{B(\boldsymbol{\alpha})}\left(1 - \sum_{i=1}^{d} x_i\right)^{\alpha_0 - 1} \prod_{i=1}^{d} x_i^{\alpha_i - 1}$$

where $B(\boldsymbol{\alpha})$ is the multinomial beta function:

$$B(\boldsymbol{\alpha}) = \frac{\prod_{i=0}^{d} \Gamma(\alpha_i)}{\Gamma\left(\sum_{i=0}^{d} \alpha_i\right)}$$

The special case of the Dirichlet distribution is the beta distribution when $d = 1$.

Marginal PDF

Let $\quad \mathbf{X} = \begin{bmatrix} \mathbf{U} \\ \mathbf{V} \end{bmatrix} \sim Dir_d(\boldsymbol{\alpha})$

Where
$$\mathbf{X} = [X_1, ..., X_s, X_{s+1}, ..., X_d]^T$$
$$\mathbf{U} = [X_1, ..., X_s]^T$$
$$\mathbf{V} = [X_{s+1}, ..., X_d]^T$$

Let $\quad \alpha_\Sigma = \sum_{j=0}^{d} \alpha_j = $ sum of α matrix elements.

$$\mathbf{U} \sim Dir_s(\boldsymbol{\alpha_u}) \quad \text{where} \quad \boldsymbol{\alpha_u} = \left[\alpha_1, \alpha_2, ..., \alpha_s, \alpha_\Sigma - \sum_{j=1}^{s} \alpha_j\right]^T$$

$$f(\mathbf{u}) = \frac{\Gamma(\alpha_\Sigma)}{\Gamma\left(\alpha_\Sigma - \sum_{j=1}^{s} \alpha_j\right) \prod_{i=1}^{s} \Gamma(\alpha_i)}\left(1 - \sum_{i=1}^{s} x_i\right)^{\alpha_\Sigma - 1 - \sum_{j=1}^{s} \alpha_j} \prod_{i=1}^{s} x_i^{\alpha_i - 1}$$

When marginalized to one variable:
$$X_i \sim Beta(\alpha_i, \alpha_\Sigma - \alpha_i)$$

$$f(x_i) = \frac{\Gamma(\alpha_\Sigma)}{\Gamma(\alpha_\Sigma - \alpha_i)\Gamma(\alpha_i)} (1 - x_i)^{\alpha_\Sigma - \alpha_i - 1} x_i^{\alpha_i - 1}$$

$U|V = v \sim Dir_{d-s}(\alpha_{u|v})$ where $\alpha_{u|v} = [\alpha_{s+1}, \alpha_{s+2}, ..., \alpha_m, \alpha_0]^T$
(Kotz et al. 2000, p.488)

Conditional PDF

$$f(u|v) = \frac{\Gamma(\sum_{j=0}^{s} \alpha_j)}{\prod_{i=0}^{s} \Gamma(\alpha_i)} \left(1 - \sum_{i=1}^{s} x_i\right)^{\alpha_0 - 1} \prod_{i=1}^{s} x_i^{\alpha_i - 1}$$

CDF

$$F(\mathbf{x}) = P(X_1 \leq x_1, X_2 \leq x_2, ..., X_d \leq x_d)$$
$$= \int_0^{x_1} \int_0^{x_2} \cdots \int_0^{x_d} \left(1 - \sum_{i=1}^{d} x_i\right)^{\alpha_0 - 1} \prod_{i=1}^{d} x_i^{\alpha_i - 1} dd, ..., dx_2, dx_1$$

Numerical methods have been explored to evaluate this integral, see (Kotz et al. 2000, pp.497-500)

Reliability

$$R(\mathbf{x}) = P(X_1 > x_1, X_2 > x_2, ..., X_d > x_d)$$
$$= \int_{x_1}^{\infty} \int_{x_2}^{\infty} \cdots \int_{x_d}^{\infty} \left(1 - \sum_{i=1}^{d} x_i\right)^{\alpha_0 - 1} \prod_{i=1}^{d} x_i^{\alpha_i - 1} dd, ..., dx_2, dx_1$$

Properties and Moments

Median   Solve numerically using $F(x) = 0.5$

Mode $x_i = \frac{\alpha_i - 1}{\alpha_\Sigma - d}$ for $\alpha_i > 0$ otherwise no mode

Mean - 1st Raw Moment   Let $\alpha_\Sigma = \sum_{i=0}^{d} \alpha_i$:

$$E[X] = \mu = \frac{\alpha}{\alpha_\Sigma}$$

Mean of the marginal distribution:
$$E[U] = \mu_u = \frac{\alpha_u}{\alpha_\Sigma}$$

$$E[X_i] = \mu_i = \frac{\alpha_i}{\alpha_\Sigma}$$

where

$$\alpha_u = \left[\alpha_1, \alpha_2, ..., \alpha_s, \alpha_\Sigma - \sum_{j=1}^{s} \alpha_j\right]^T$$

Mean of the conditional distribution:
$$E[U|V = v] = \mu_{u|v} = \frac{\alpha_{u|v}}{\alpha_\Sigma}$$

where
$$\alpha_{u|v} = [\alpha_{s+1}, \alpha_{s+2}, ..., \alpha_m, \alpha_0]^T$$

Variance - 2nd Central Moment   Let $\alpha_\Sigma = \sum_{i=0}^{d} \alpha_i$:

$$Var[X_i] = \frac{\alpha_i(\alpha_\Sigma - \alpha_i)}{\alpha_\Sigma^2(\alpha_\Sigma + 1)}$$

$$Cov[X_i, X_j] = \frac{-\alpha_i \alpha_j}{\alpha_\Sigma^2(\alpha_\Sigma + 1)}$$

Parameter Estimation

Maximum Likelihood Function

MLE Point Estimates

The MLE estimates of $\hat{\alpha}_i$ can be obtained from n observations of x_i by numerically maximizing the log-likelihood function: (Kotz et al. 2000, p.505)

$$\Lambda(\alpha|E) = n\left\{ ln\Gamma(\alpha_\Sigma) - \sum_{j=0}^{d} \ln\Gamma(\alpha_j) \right\} + n\sum_{j=0}^{d} \left\{ \frac{1}{n}(\alpha_j - 1)\sum_{i=1}^{n} \ln(x_{ij}) \right\}$$

The method of moments is used to provide initial guesses of α_i for the numerical methods.

Fisher Information Matrix

$$I_{ij} = -n\psi'(\alpha_\Sigma), \qquad i \neq j$$
$$I_{ii} = n\psi'(\alpha_i) - n\psi'(\alpha_\Sigma)$$

Where $\psi'(x) = \frac{d^2}{dx^2} ln\Gamma(x)$ is the trigamma function. See Section 1.6.8. (Kotz et al. 2000, p.506)

100γ% Confidence Intervals

The confidence intervals can be obtained from the fisher information matrix.

Bayesian

Non-informative Priors

Jeffery's Prior

$$\sqrt{\det(I(\alpha))}, \text{ where } I(\alpha) \text{ is given above.}$$

Conjugate Priors

UOI	Likelihood Model	Evidence	Dist. of UOI	Prior Para	Posterior Parameters
p from $MNom_d(k; n_t, p)$	Multinomial$_d$	$k_{i,j}$ failures in n trials with d possible states.	Dirichlet$_{d+1}$	α_o	$\alpha = \alpha_o + k$

Description, Limitations and Uses

Example

Five machines are measured for performance on demand. The machines can either fail, partially fail or success in their application. The machines are tested for 10 demands with the following data for each machine:

Machine/Trail	1	2	3	4	5	6	7	8	9	10
1		F = 3		P = 2				S = 5		
2		F=2		P=2				S=6		
3		F=2		P=3				S=5		
4		F=3		P=3				S=4		
5		F=2		P=3				S=5		
μ_i		$n\widehat{p_F}$		$n\widehat{p_P}$				$n\widehat{p_F}$		

Estimate the multinomial distribution parameter $p = [p_F, p_P, p_S]$:

Using a non-informative improper prior $Dir_3(0,0,0)$ after updating:

$$x = \begin{bmatrix} p_F \\ p_P \\ p_S \end{bmatrix} \quad \alpha = \begin{bmatrix} 12 \\ 13 \\ 25 \end{bmatrix} \quad E[x] = \begin{bmatrix} \widehat{p_F} = \frac{12}{50} \\ \widehat{p_P} = \frac{13}{50} \\ \widehat{p_S} = \frac{25}{50} \end{bmatrix} \quad Var[x] = \begin{bmatrix} 7.15E-5 \\ 7.54E-5 \\ 9.80E-5 \end{bmatrix}$$

Confidence intervals for the parameters $p = [p_F, p_P, p_S]$ can also be calculated using the cdf of the marginal distribution $F(x_i)$.

Characteristic

Beta Generalization. The Dirichlet distribution is a generalization of the beta distribution. The beta distribution is seen when $d = 1$.

α **Interpretation.** The higher α_i the sharper and more certain the distribution is. This follows from its use in Bayesian statistics to model the multinomial distribution parameter p. As more evidence is used, the α_i values get higher which reduces uncertainty. The values of α_i can also be interpreted as a count for each state of the multinomial distribution.

Alternative Formulation. The most common formulation of the Dirichlet distribution is as follows:
$$\alpha = [\alpha_1, \alpha_2, ..., \alpha_m]^T \quad \text{where} \quad \alpha_i > 0$$
$$x = [x_1, x_2, ..., x_m]^T \quad \text{where} \quad 0 \le x_i \le 1, \quad \sum_{i=1}^{m} x_i = 1$$
$$f(x) = \frac{1}{B(\alpha)} \prod_{i=1}^{m} x_i^{\alpha_i - 1}$$

This formulation is popular because it is a simpler presentation where the matrix of α and x are the same size. However, it should be noted that last term of the vector x is dependent on $\{x_1 ... x_{m-1}\}$ through the relationship $x_m = 1 - \sum_{i=1}^{m-1} x_i$.

Neutrality. (Kotz et al. 2000, p.500) If X_1 and X_2 are non-negative random variables such that $X_1 + X_2 \le 1$ then X_i is called neutral if the following are independent:

$$X_i \perp \frac{X_j}{1 - X_i} \quad (i \ne j)$$

If $X \sim Dir_d(\alpha)$ then X is a neutral vector with each X_i being neutral under all permutations of the above definition. This property is unique to the Dirichlet distribution.

Applications   **Bayesian Statistics.**   The Dirichlet distribution is often used as a conjugate prior to the multinomial likelihood function.

Resources   Online:
http://en.wikipedia.org/wiki/Dirichlet_distribution
http://www.cis.hut.fi/ahonkela/dippa/node95.html

Books:
Kotz, S., Balakrishnan, N. & Johnson, N.L., 2000. *Continuous Multivariate Distributions, Volume 1, Models and Applications*, 2nd Edition 2nd ed., Wiley-Interscience.

Congdon, P., 2007. Bayesian Statistical Modelling 2nd ed., Wiley.

MacKay, D.J. & Petoy, L.C., 1995. *A hierarchical Dirichlet language model*. Natural language engineering.

Relationship to Other Distributions

Beta
Distribution

$Beta(x; \alpha, \beta)$

Special Case:
$$Dir_{d=1}(x; [\alpha_1, \alpha_0]) = Beta(k = x; \alpha = \alpha_1, \beta = \alpha_0)$$

Gamma
Distribution

$Gamma(x; \lambda, k)$

Let:
$$Y_i \sim Gamma(\lambda, k_i) \ \ i.i.d \ and \ \ V = \sum_{i=1}^{d} Y_i$$
Then:
$$V \sim Gamma(\lambda, \textstyle\sum k_i)$$
Let:
$$Z = \left[\frac{Y_1}{V}, \frac{Y_2}{V}, ..., \frac{Y_d}{V}\right]$$
Then:
$$Z \sim Dir_d(\alpha_1, ..., \alpha_k)$$

*i.i.d: independent and identically distributed

6.3. Multivariate Normal Continuous Distribution

*Note for a graphical representation see bivariate normal distribution

Parameters & Description

Parameters			
	$\boldsymbol{\mu} = [\mu_1, \mu_2, \dots, \mu_d]^T$	$-\infty < \mu_i < \infty$	*Location Vector:* A d-dimensional vector giving the mean of each random variable.
	$\Sigma = \begin{bmatrix} \sigma_{11} & \cdots & \sigma_{1d} \\ \vdots & \ddots & \vdots \\ \sigma_{d1} & \cdots & \sigma_{dd} \end{bmatrix}$	$\sigma_{ii} > 0$ $\sigma_{ij} \geq 0$	*Covariance Matrix:* A $d \times d$ matrix which quantifies the random variable variance and dependence. This matrix determines the shape of the distribution. Σ is symmetric positive definite matrix.
	d	$d \geq 2$ (integer)	*Dimensions.* The number of dependent variables.

Limits $-\infty < x_i < \infty$

Distribution   Formulas

PDF

$$f(\mathbf{x}) = \frac{1}{(2\pi)^{d/2}\sqrt{|\Sigma|}} \exp\left[-\frac{1}{2}(\mathbf{x} - \boldsymbol{\mu})^T \Sigma^{-1}(\mathbf{x} - \boldsymbol{\mu})\right]$$

Where $|\Sigma|$ is the determinant of Σ.

Let $\qquad X = \begin{bmatrix} U \\ V \end{bmatrix} \sim Norm_d\left(\begin{bmatrix} \mu_u \\ \mu_v \end{bmatrix}, \begin{bmatrix} \Sigma_{uu} & \Sigma_{uv} \\ \Sigma_{uv}^T & \Sigma_{vv} \end{bmatrix}\right)$

Where $\qquad X = [X_1, \dots, X_p, X_{p+1}, \dots, X_d]^T$
$\qquad\qquad U = [X_1, \dots, X_p]^T$
$\qquad\qquad V = [X_{p+1}, \dots, X_d]^T$

Marginal PDF

$$U \sim Norm_p(\boldsymbol{\mu_u}, \Sigma_{uu})$$

$$f(\boldsymbol{u}) = \int_{-\infty}^{\infty} f(\boldsymbol{x})\, dv$$

$$= \frac{1}{(2\pi)^{p/2}\sqrt{|\Sigma_{uu}|}} \exp\left[-\frac{1}{2}(\boldsymbol{u} - \boldsymbol{\mu_u})^T \Sigma_{uu}^{-1}(\boldsymbol{u} - \boldsymbol{\mu_u})\right]$$

$$U|V = v \sim Norm_p(\boldsymbol{\mu_{u|v}}, \Sigma_{u|v})$$

Conditional PDF   Where

$$\boldsymbol{\mu_{u|v}} = \boldsymbol{\mu_u} + \Sigma_{uv}^T \Sigma_{vv}^{-1}(\boldsymbol{v} - \boldsymbol{\mu_v})$$
$$\Sigma_{u|v} = \Sigma_{uu} - \Sigma_{uv}^T \Sigma_{vv}^{-1} \Sigma_{uv}$$

CDF	$F(\mathbf{x}) = \dfrac{1}{(2\pi)^{d/2}\sqrt{\vert\Sigma\vert}} \displaystyle\int_{-\infty}^{\mathbf{x}} \exp\left[-\dfrac{1}{2}(\mathbf{x}-\boldsymbol{\mu})^{\mathrm{T}}\Sigma^{-1}(\mathbf{x}-\boldsymbol{\mu})\right] d\mathbf{x}$
Reliability	$R(\mathbf{x}) = \dfrac{1}{(2\pi)^{d/2}\sqrt{\vert\Sigma\vert}} \displaystyle\int_{\mathbf{x}}^{\infty} \exp\left[-\dfrac{1}{2}(\mathbf{x}-\boldsymbol{\mu})^{\mathrm{T}}\Sigma^{-1}(\mathbf{x}-\boldsymbol{\mu})\right] d\mathbf{x}$

Properties and Moments

Median	$\boldsymbol{\mu}$
Mode	$\boldsymbol{\mu}$
Mean - 1st Raw Moment	$E[X] = \boldsymbol{\mu}$
	Mean of the marginal distribution: $$E[U] = \boldsymbol{\mu}_u$$ $$E[V] = \boldsymbol{\mu}_v$$
	Mean of the conditional distribution: $$\boldsymbol{\mu}_{u\vert v} = \boldsymbol{\mu}_u + \Sigma_{uv}^{T}\Sigma_{vv}^{-1}(v - \boldsymbol{\mu}_v)$$
Variance - 2nd Central Moment	$Cov[X] = \Sigma$
	Covariance of marginal distributions: $$Cov(\mathbf{U}) = \Sigma_{uu}$$
	Covariance of conditional distributions: $$Cov(\mathbf{U}\vert\mathbf{V}) = \Sigma_{uu} - \Sigma_{uv}^{T}\Sigma_{vv}^{-1}\Sigma_{uv}$$

Parameter Estimation

Maximum Likelihood Function

MLE Point Estimates

When given complete data of n_F samples:

$$x_t = \left[x_{1,t}, x_{2,t}, \dots, x_{d,t}\right]^{T} \quad \text{where } t = (1, 2, \dots, n_F)$$

The following MLE estimates are given: (Kotz et al. 2000, p.161)

$$\hat{\boldsymbol{\mu}} = \frac{1}{n_F}\sum_{t=1}^{n_F} x_t$$

$$\hat{\Sigma}_{ij} = \frac{1}{n_F}\sum_{t=1}^{n_F}\left(x_{i,t} - \hat{\mu}_i\right)\left(x_{j,t} - \hat{\mu}_j\right)$$

A review of different estimators is given in (Kotz et al. 2000). When estimates are from a low number of samples ($n_F < 30$) a correction factor of -1 can be introduced to give the unbiased estimators (Tong 1990, p.53):

$$\hat{\Sigma}_{ij} = \frac{1}{n_F - 1}\sum_{t=1}^{n_F}\left(x_{i,t} - \hat{\mu}_i\right)\left(x_{j,t} - \hat{\mu}_j\right)$$

Fisher Information Matrix		$I_{i,j} = \dfrac{\partial \mu^T}{\partial \theta_i} \Sigma^{-1} \dfrac{\partial \mu}{\partial \theta_j}$

Bayesian

Non-informative Priors when Σ is known, $\pi_0(\mu)$
(Yang and Berger 1998, p.22)

Type	Prior	Posterior
Uniform Improper, Jeffrey, Reference Prior	1	$\pi(\mu\|E) \sim Norm_d\left(\mu; \dfrac{1}{n_F}\sum_{t=1}^{n_F} x_t, \dfrac{\Sigma}{n_F}\right)$ when $\mu \in (\infty, \infty)$
Shrinkage	$(\mu^T \Sigma^{-1} \mu^T)^{-(d-2)}$	No Closed Form

Non-informative Priors when μ is known, $\pi_o(\Sigma)$
(Yang & Berger 1994)

Type	Prior	Posterior
Uniform Improper Prior with limits $\Sigma \in (0, \infty)$	1	$\pi(\Sigma^{-1}\|E) \sim$ $Wishart_d\left(\Sigma^{-1}; n_F - d - 1, \dfrac{S^{-1}}{n_F}\right)$
Jeffery's Prior	$\dfrac{1}{\|\Sigma\|^{\frac{d+1}{2}}}$	$\pi(\Sigma^{-1}\|E) \sim$ $Wishart_d\left(\Sigma^{-1}; n_F, \dfrac{S^{-1}}{n_F}\right)$ with limits $\Sigma \in (0, \infty)$
Reference Prior Ordered $\{\lambda_i, \lambda_j, .., \lambda_d\}$	$\dfrac{1}{\|\Sigma\| \prod_{i<j}(\lambda_i - \lambda_j)}$	Proper - No Closed Form
Reference Prior Ordered $\{\lambda_1, \lambda_d, \lambda_i, .., \lambda_{d-1}\}$	$\dfrac{1}{\|\Sigma\|(\log\lambda_1 - \log\lambda_d)^{d-2} \prod_{i<j}(\lambda_i - \lambda_j)}$	Proper - No Closed Form
MDIP	$\dfrac{1}{\|\Sigma\|}$	No Closed Form

Non-informative Priors when μ and Σ are unknown for bivariate normal, $\pi_o(\mu, \Sigma)$. A complete coverage of numerous reference prior distributions with different parameter ordering is contained in (Berger & Sun 2008)

Type	Prior	Posterior
Uniform Improper Prior	1	No Closed Form
Jeffery's Prior	$\dfrac{1}{\|\Sigma\|^{\frac{d+1}{2}}}$	No Closed Form

Reference Prior Ordered $\{\lambda_i, \lambda_j, .., \lambda_d\}$	$\dfrac{1}{\|\Sigma\| \prod_{i<j}(\lambda_i - \lambda_j)}$	No Closed Form
Reference Prior Ordered $\{\lambda_1, \lambda_d, \lambda_i, .., \lambda_{d-1}\}$	$\dfrac{1}{\|\Sigma\|(\log\lambda_1 - \log\lambda_d)^{d-2} \prod_{i<j}(\lambda_i - \lambda_j)}$	No Closed Form
MDIP	$\dfrac{1}{\|\Sigma\|}$	No Closed Form

Where, λ_i is the ith eigenvalue of Σ, and $\bar{R}$ and R are population and sample multiple correlation coefficients. Where,

$$S_{ij} = \frac{1}{n_F - 1}\sum_{t=1}^{n_F}(x_{i,t} - \hat{\mu}_i)(x_{j,t} - \hat{\mu}_j) \quad \text{and} \quad \bar{x} = \frac{1}{n_F}\sum_{t=1}^{n_F} x_t$$

Conjugate Priors

UOI	Likelihood Model	Evidence	Dist. of UOI	Prior Para	Posterior Parameters
μ from $Norm_d(\mu, \Sigma)$	Multi-variate Normal with known Σ	n_F events at x points	Multi-variate Normal	U_0, V_0	$U = \dfrac{V_0^{-1}U_o + n_F V^{-1}\bar{x}}{V_0^{-1} + n_F \Sigma^{-1}}$ $V = \dfrac{1}{V_0^{-1} + n_F \Sigma^{-1}}$

Description, Limitations and Uses

Example	See bivariate normal distribution.
Characteristic	**Standard Spherical Normal Distribution.** When $\mu = 0$, $\Sigma = I$ we obtain the standard spherical normal distribution:

$$f(\mathbf{x}) = \frac{1}{(2\pi)^{d/2}} \exp\left[-\frac{1}{2}\mathbf{x}^T\mathbf{x}\right]$$

Covariance Matrix. (Yang et al. 2004, p.49)
- **Diagonal Elements.** The diagonal elements of Σ is the variance of each random variable. $\sigma_{ii} = Var(X_i)$
- **Non-Diagonal Elements.** Non-diagonal elements give the covariance $\sigma_{ij} = Cov(X_i, X_j) = \sigma_{ji}$. Hence the matrix is symmetric.
- **Independent Variables.** If $Cov(X_i, X_j) = \sigma_{ij} = 0$ then X_i and X_j and independent.
- $\sigma_{ij} > 0$. When X_i increases then X_j and tends to increase.
- $\sigma_{ij} < 0$. When X_i increases then X_j and tends to decrease.

Ellipsoid Axis. The ellipsoid has axes pointing in the direction of the eigenvectors of Σ. A magnitude of these axes is given by the corresponding eigenvalues.

Mean / Median / Mode:
As per the univariate distributions the mean, median and mode are equal.

Convolution Property
Let $X \sim Norm_d(\mu_x, \Sigma_x)$ $Y \sim Norm_d(\mu_y, \Sigma_y)$
Where $X \perp Y$ (independent)
Then $X + Y \sim Norm_d(\mu_x + \mu_y, \Sigma_x + \Sigma_y)$

Note, if **X** and **Y** are dependent then **X** + **Y** may not be normally distributed. (Novosyolov 2006)

Scaling Property
Let $Y = AX + b$ Y is a p x 1 matrix
                                                 b is a p x 1 matrix

Then $Y \sim Norm_d(A\mu + b, A\Sigma A^T)$ A is a p x d matrix

Marginalize Property:
Let $X = \begin{bmatrix} U \\ V \end{bmatrix} \sim Norm_d\left(\begin{bmatrix} \mu_u \\ \mu_v \end{bmatrix}, \begin{bmatrix} \Sigma_{uu} & \Sigma_{uv} \\ \Sigma_{uv}^T & \Sigma_{vv} \end{bmatrix}\right)$

Then $U \sim Norm_p(\mu_u, \Sigma_{uu})$ U is a p x 1 matrix

Conditional Property:
Let $X = \begin{bmatrix} U \\ V \end{bmatrix} \sim Norm_d\left(\begin{bmatrix} \mu_u \\ \mu_v \end{bmatrix}, \begin{bmatrix} \Sigma_{uu} & \Sigma_{uv} \\ \Sigma_{uv}^T & \Sigma_{vv} \end{bmatrix}\right)$

Then $U|V = v \sim Norm_p(\mu_{u|v}, \Sigma_{u|v})$ U is a p x 1 matrix

Where $\mu_{u|v} = \mu_u + \Sigma_{uv}^T \Sigma_{vv}^{-1}(V - \mu_v)$
                  $\Sigma_{u|v} = \Sigma_{uu} - \Sigma_{uv}^T \Sigma_{vv}^{-1} \Sigma_{uv}$

It should be noted that the standard deviation of the marginal distribution does not depend on the given values in **V**.

Applications   **Convenient Properties.** (Balakrishnan & Lai 2009, p.477) Popularity of the multivariate normal distribution over other multivariate distributions is due to the convenience of the conditional and marginal distribution properties which both produce univariate normal distributions.

Kalman Filter. The Kalman filter estimates the current state of a system in the presence of noisy measurements. This process uses multivariate normal distributions to model the noise.

Multivariate Analysis of Variance (MANOVA). A test used to analyze variance and dependence of variables. A popular model used to conduct MANOVA assumes the data comes from a multivariate normal population.

Gaussian Regression Process. This is a statistical model for observations or events that occur in a continuous domain of time or space, where every point is associated with a normally distributed random variable and every finite collection of these random variables has a multivariate normal distribution.

Multi-Linear Regression. Multi-linear regression attempts to model the relationship between parameters and variables by fitting a linear equation. One model to do such a task (MLE) fits a distribution to the observed variance where a multivariate normal distribution is often assumed.

Gaussian Bayesian Belief Networks (BBN). BBNs graphical represent the dependence between variables in a probability distribution. When using continuous random variables BBNs quickly become tremendously complicated. However due to the multivariate normal distribution's conditional and marginal properties this task is simplified and popular.

Resources   Online:
http://mathworld.wolfram.com/BivariateNormalDistribution.html
http://www.aiaccess.net/English/Glossaries/GlosMod/e_gm_binormal _distri.htm (interactive visual representation)

Books:
Patel, J.K, Read, C.B, 1996. *Handbook of the Normal Distribution*, 2nd Edition, CRC

Tong, Y.L., 1990. *The Multivariate Normal Distribution*, Springer.

Yang, K. et al., 2004. *Multivariate Statistical Methods in Quality Management* 1st ed., McGraw-Hill Professional.

Bertsekas, D.P. & Tsitsiklis, J.N., 2008. *Introduction to Probability*, 2nd Edition, Athena Scientific.

6.4. **Multinomial Discrete Distribution**

Probability Density Function - f(**k**)

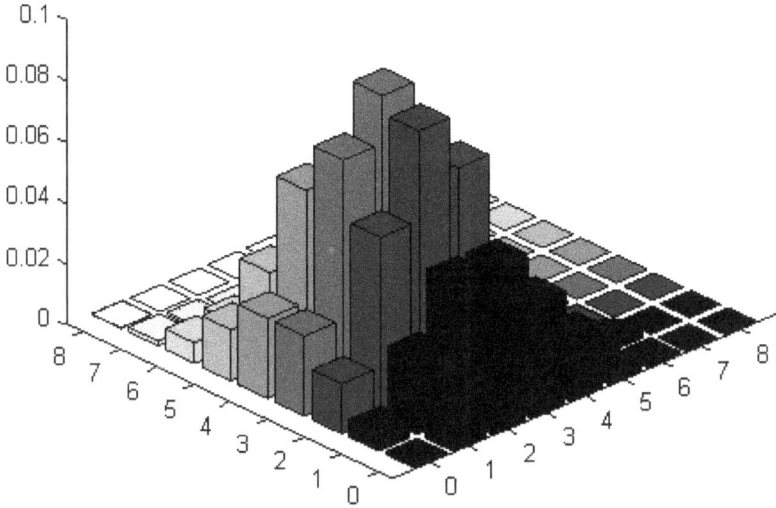

Trinomial Distribution, $f([k_1, k_2, k_3]^T)$ where $n = 8$, $\mathbf{p} = \left[\frac{1}{3}, \frac{1}{4}, \frac{5}{12}\right]^T$. Note k_3 is not shown because it is determined using $k_3 = n - k_1 - k_2$

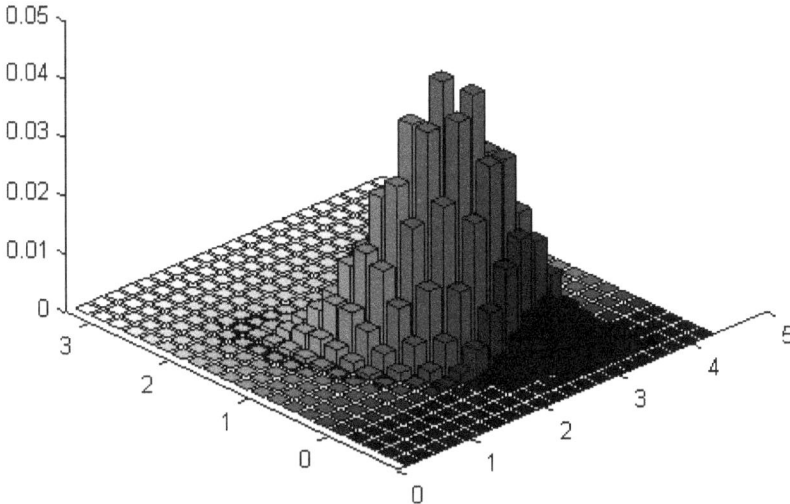

Trinomial Distribution, $f([k_1, k_2, k_3]^T)$ where $n = 20$, $\mathbf{p} = \left[\frac{1}{3}, \frac{1}{2}, \frac{1}{6}\right]^T$. Note k_3 is not shown because it is determined as $k_3 = n - k_1 - k_2$

Parameters & Description

	n	$n > 0$ (integer)	*Number of Trials.* This is sometimes called the index. (Johnson et al. 1997, p.31)
Parameters	$\mathbf{p} = [p_1, p_2, \ldots, p_d]^T$	$0 \leq p_i \leq 1$ $\sum_{i=1}^{d} p_i = 1$	*Event Probability Matrix:* The probability of event *i* occurring. p_i is often called cell probabilities. (Johnson et al. 1997, p.31)
	d	$d \geq 2$ (integer)	*Dimensions.* The number of mutually exclusive states of the system.

Limits	$k_i \in \{0, \ldots, n\}$ $$\sum_{i=1}^{d} k_i = n$$

Distribution	Formulas

PDF

$$f(\mathbf{k}) = \binom{n}{k_1, k_2, \ldots, k_d} \prod_{i=1}^{d} p_i^{k_i}$$

where

$$\binom{n}{k_1, k_2, \ldots, k_n} = \frac{n!}{k_1! k_2! \ldots k_d!} = \frac{n!}{\prod_{i=1}^{d} k_i!} = \frac{\Gamma(n+1)}{\prod_{i=1}^{d} \Gamma(k_i + 1)}$$

Note that in p there is only d-1 'free' variables as the last $p_d = 1 - \sum_{i=1}^{d-1} p_i$ giving the distribution:

$$f(\mathbf{k}) = \binom{n}{k_1, k_2, \ldots, k_n} \prod_{i=1}^{d-1} p_i^{k_i} \cdot \left(1 - \sum_{i=1}^{s} p_i\right)^{n - \sum_{i=1}^{d-1} k_i}$$

Now the special case of binomial distribution when $d = 2$ can be seen.

Marginal PDF

Let $\quad K = \begin{bmatrix} U \\ V \end{bmatrix} \sim MNom_d \left(n, \begin{bmatrix} \boldsymbol{p_u} \\ \boldsymbol{p_v} \end{bmatrix}\right)$

Where $\quad K = [K_1, \ldots, K_s, K_{s+1}, \ldots, K_d]^T$
$\quad U = [K_1, \ldots, K_s]^T$
$\quad V = [K_{s+1}, \ldots, K_d]^T$

$$U \sim MNom_s(n, \boldsymbol{p_u})$$

where $\quad \boldsymbol{p_u} = [p_1, p_2, \ldots, p_{s-1}, (1 - \sum_{i=1}^{s-1} p_i)]^T$

$$f(\boldsymbol{u}) = \binom{n}{k_1, k_2, \ldots, k_s} \prod_{i=1}^{s} p_i^{k_i}$$

When only two states $\boldsymbol{p} = [p, (1 - p)]^T$:

$$f(k_i) = \binom{n}{k_i} p_i^{k_i} (1 - p_i)^{n - k_i}$$

$$\boldsymbol{U}|\boldsymbol{V} = \boldsymbol{v} \sim MNom_s\left(n_{u|v}, \boldsymbol{p}_{u|v}\right)$$

where

Conditional PDF

$$n_{u|v} = n - n_v = n - \sum_{i=s+1}^{d} k_i$$

$$\boldsymbol{p}_{u|v} = \frac{1}{\sum_{i=1}^{s} p_i} [p_1, p_2, \dots, p_s]^T$$

CDF

$$F(\mathbf{k}) = P(K_1 \leq k_1, K_2 \leq k_2, \dots, K_d \leq k_d)$$

$$= \sum_{j_1=0}^{k_1} \sum_{j_2=0}^{k_2} \cdots \sum_{j_d=0}^{k_d} \binom{n}{j_1, j_2, \dots, j_d} \prod_{i=1}^{d} p_i^{j_i}$$

Reliability

$$R(\mathbf{k}) = P(K_1 > k_1, K_2 > k_2, \dots, K_d > k_d)$$

$$= \sum_{j_1=k_1+1}^{n} \sum_{j_2=k_2+1}^{n} \cdots \sum_{j_d=k_d+1}^{n} \binom{n}{j_1, j_2, \dots, j_d} \prod_{i=1}^{d} p_i^{j_i}$$

Properties and Moments

Median[4]

$$Median(k_i) \text{ is either } \{\lfloor np_i \rfloor, \lceil np_i \rceil\}$$

Mode

$$Mode(k_i) = \lfloor (n + 1)p_i \rfloor$$

Mean - 1st Raw Moment

$$E[\boldsymbol{K}] = \boldsymbol{\mu} = n\boldsymbol{p}$$

Mean of the marginal distribution:
$$E[\boldsymbol{U}] = \boldsymbol{\mu}_u = n\boldsymbol{p}_u$$
$$E[K_i] = \mu_{k_i} = np_i$$

Mean of the conditional distribution:
$$E[\boldsymbol{U}|\boldsymbol{V} = \boldsymbol{v}] = \boldsymbol{\mu}_{u|v} = n_{u|v}\boldsymbol{p}_{u|v}$$
where

$$n_{u|v} = n - n_v = n - \sum_{i=s+1}^{d} k_i$$

$$\boldsymbol{p}_{u|v} = \frac{1}{\sum_{i=1}^{s} p_i} [p_1, p_2, \dots, p_s]^T$$

Variance - 2nd Central Moment

$$Var[K_i] = np_i(1 - p_i)$$
$$Cov[K_i, K_j] = -np_i p_j$$

Covariance of marginal distributions:
$$Var[K_i] = np_i(1 - p_i)$$

Covariance of conditional distributions:

---

[4] $\lfloor x \rfloor$ = is the floor function (largest integer not greater than x)
$\lceil x \rceil$ = is the ceiling function (smallest integer not less than x)

$$Var[K_{U|V,i}] = n_{u|v}p_{u|v,i}(1 - p_{u|v,i})$$
$$Cov[K_{U|V,i}, K_{U|V,j}] = -n_{u|v}p_{u|v,i}p_{u|v,j}$$

where

$$n_{u|v} = n - n_v = n - \sum_{i=s+1}^{d} k_i$$

$$\mathbf{p}_{u|v} = \frac{1}{\sum_{i=1}^{s} p_i}[p_1, p_2, \dots, p_s]^T$$

Parameter Estimation

Maximum Likelihood Function

MLE Point Estimates

As with the binomial distribution the MLE estimates, given the vector $\mathbf{k}$(and therefore n), is:(Johnson et al. 1997, p.51)

$$\hat{\mathbf{p}} = \frac{\mathbf{k}}{n}$$

Where there are T observations of $\mathbf{k}_t$ each containing n_t trails:

$$\hat{\mathbf{p}} = \frac{1}{\sum_{t=1}^{n} n_t}\sum_{t=1}^{T} \mathbf{k}_t$$

$100\gamma\%$ Confidence Intervals

(Complete Data)

An approximation of the joint interval confidence limits for $100\gamma\%$ given by Goodman in 1965 is:(Johnson et al. 1997, p.51)

p_i lower confidence limit:

$$\frac{1}{2(n + A)}\left[A + 2k_i - A\sqrt{A + \frac{4}{n}k_i(n - k_i)}\right]$$

p_i upper confidence limit:

$$\frac{1}{2(n + A)}\left[A + 2k_i + A\sqrt{A + \frac{4}{n}k_i(n - k_i)}\right]$$

where Φ is the standard normal CDF and:

$$A = Z_{\frac{d-1+\gamma}{d}} = \Phi^{-1}\left(\frac{d - 1 + \gamma}{d}\right)$$

A complete coverage of estimation techniques and confidence intervals is contained in (Johnson et al. 1997, pp.51-65). A more accurate method which requires numerical methods is given in (Sison & Glaz 1995)

Bayesian

Non-informative Priors, $\pi(p)$
(Yang and Berger 1998, p.6)

Type	Prior	Posterior
Uniform Prior	$1 = Dir_{d+1}(\alpha_i = 1)$	$Dir_{d+1}\left(\mathbf{p}\mid 1 + \mathbf{k}\right)$
Jeffreys Prior One Group - Reference Prior	$\dfrac{C}{\sqrt{\prod_{i=1}^{d} p_i}} = Dir_{d+1}\left(\alpha_i = \frac{1}{2}\right)$	$Dir_{d+1}\left(\mathbf{p}\mid \frac{1}{2}+\mathbf{k}\right)$

where C is a constant

In terms of the reference prior, this approach considers all parameters are of equal importance.(Berger & Bernardo 1992)

d-group Reference Prior	$\dfrac{C}{\sqrt{\prod_{i=1}^{d-1}\{p_i(1 - \sum_{j=1}^{i} p_j)\}}}$	Proper. See m-group posterior when $m = 1$.

where C is a constant

This approach considers each parameter to be of different importance (group length 1) and so the parameters must be ordered by importance. (Berger & Bernardo 1992)

m-group Reference Prior

$$\pi_o(p) = \frac{C}{\sqrt{\left(1 - \sum_{j=1}^{N_m} p_j\right) \prod_{i=1}^{d-1} p_i \prod_{i=1}^{m-1}\left(1 - \sum_{j=1}^{N_i} p_j\right)^{n_{i+1}}}}$$

where groups are given by:

$$\mathbf{p_1} = [p_1, \dots p_{n_1}]^T, \quad \mathbf{p_2} = [p_{n_1+1}, \dots, p_{n_1+n_2}]^T$$
$$N_j = n_1 + \dots + n_j \text{ for } j = 1, \dots, m$$
$$\mathbf{p_i} = [p_{N_{i-1}+1}, \dots, p_{N_i}]^T$$
C is a constant

Posterior:

$$\pi(p|k) \propto \frac{\left(1 - \sum_{j=1}^{N_m} p_j\right)^{k_d - \frac{1}{2}}}{\sqrt{\prod_{i=1}^{d-1} p_i \prod_{i=1}^{m-1}\left(1 - \sum_{j=1}^{N_i} p_j\right)^{n_{i+1}}}}$$

This approach splits the parameters into m different groups of importance. Within the group order is not important, but the groups need to be ordered by importance. It is common to have $m = 2$ and split the parameters into importance and nuisance parameters. (Berger & Bernardo 1992)

MDIP	$\prod_{i=1}^{d} p_i^{p_i} = Dir_{d+1}(\alpha_i = p_i + 1)$	$Dir_{d+1}\left(\mathbf{p}'\mid p_i + 1 + k_i\right)$
Novick and Hall's Prior (improper)	$\prod_{i=1}^{d} p_i^{-1} = Dir_{d+1}(\alpha_i = 0)$	$Dir_{d+1}\left(\mathbf{p}\mid \mathbf{k}\right)$

Conjugate Priors (Fink 1997)

UOI	Likelihood Model	Evidence	Dist of UOI	Prior Para	Posterior Parameters
p from $MNom_d(k; n_t, p)$	Multinomial$_d$	$k_{i,j}$ failures in n trials with d possible states.	Dirichlet$_{d+1}$	α_o	$\alpha = \alpha_o + k$

Description, Limitations and Uses

Example   A six-sided dice being thrown 60 times produces the following multinomial distribution:

Face Number	Times Observed
1	12
2	7
3	11
4	10
5	8
6	12

$$k = \begin{bmatrix} 12 \\ 6 \\ 12 \\ 10 \\ 8 \\ 12 \end{bmatrix} \quad p = \begin{bmatrix} 0.2 \\ 0.1 \\ 0.2 \\ 0.1\dot{6} \\ 0.1\dot{3} \\ 0.2 \end{bmatrix} \quad n = 60$$

Characteristic   **Binomial Generalization.** The multinomial distribution is a generalization of the binomial distribution where more than two states of the system are allowed. The binomial distribution is a special case where $d = 2$.

Covariance. All covariance's are negative. This is because the increase in one parameter p_i must result in the decrease of p_j to satisfy $\Sigma p_i = 1$.

With Replacement. The multinomial distribution assumes replacement. The equivalent distribution which assumes without replacement is the multivariate hypergeometric distribution.

Convolution Property
Let
$$K_t \sim MNom_d(k; n_t, \mathbf{p})$$
Then
$$\sum K_t \sim MNom_d(\Sigma k_t; \Sigma n_t, \mathbf{p})$$

*This does not hold when the **p** parameter differs.

Applications   **Partial Failures.** When the states of a system under demands cannot be modeled with two states (success or failure) the multinomial distribution may be used. Examples of this include when modeling discrete states of component degradation.

Resources   Online:
http://en.wikipedia.org/wiki/Multinomial_distribution

http://mathworld.wolfram.com/MultinomialDistribution.html
http://www.math.uah.edu/stat/bernoulli/Multinomial.xhtml

Books:
Johnson, N.L., Kotz, S. & Balakrishnan, N., 1997. *Discrete Multivariate Distributions* 1st ed., Wiley-Interscience.

Relationship to Other Distributions

Binominal Distribution	Special Case:
$Binom(k\|n,p)$	$MNom_{d=2}(\mathbf{k}\|\mathrm{n},\mathbf{p}) = Binom(k\|n,p)$

7. References

Abadir, K. & Magdalinos, T., 2002. The Characteristic Function from a Family of Truncated Normal Distributions. *Econometric Theory*, 18(5), p.1276-1287.

Agresti, A., 2002. *Categorical data analysis*, John Wiley and Sons.

Aitchison, J.J. & Brown, J.A.C., 1957. *The Lognormal Distribution*, New York: Cambridge University Press.

Andersen, P.K. et al., 1996. *Statistical Models Based on Counting Processes* Corrected., Springer.

Angus, J.E., 1994. Bootstrap one-sided confidence intervals for the log-normal mean. *Journal of the Royal Statistical Society. Series D (The Statistician)*, 43(3), p.395–401.

Anon, Six Sigma | Process Management | Strategic Process Management | Welcome to SSA & Company.

Antle, C., Klimko, L. & Harkness, W., 1970. Confidence Intervals for the Parameters of the Logistic Distribution. *Biometrika*, 57(2), p.397-402.

Aoshima, M. & Govindarajulu, Z., 2002. Fixed-width confidence interval for a lognormal mean. *International Journal of Mathematics and Mathematical Sciences*, 29(3), p.143–153.

Arnold, B.C., 1983. *Pareto distributions*, Fairland, MD: International Co-operative Pub. House.

Artin, E., 1964. *The Gamma Function*, New York: Holt, Rinehart & Winston.

Balakrishnan, 1991. *Handbook of the Logistic Distribution* 1st ed., CRC.

Balakrishnan, N. & Basu, A.P., 1996. *Exponential Distribution: Theory, Methods and Applications* 1st ed., CRC.

Balakrishnan, N. & Lai, C.-D., 2009. *Continuous Bivariate Distributions* 2nd ed., Springer.

Balakrishnan, N. & Rao, C.R., 2001. *Handbook of Statistics 20: Advances in Reliability* 1st ed., Elsevier Science & Technology.

Berger, J.O., 1993. *Statistical Decision Theory and Bayesian Analysis* 2nd ed., Springer.

Berger, J.O. & Bernardo, J.M., 1992. Ordered Group Reference Priors with Application to the Multinomial Problem. *Biometrika*, 79(1), p.25-37.

Berger, J.O. & Sellke, T., 1987. Testing a Point Null Hypothesis: The Irreconcilability of P Values and Evidence. *Journal of the American Statistical Association*, 82(397), p.112-122.

Berger, J.O. & Sun, D., 2008. Objective priors for the bivariate normal model. *The Annals of Statistics*, 36(2), p.963-982.

Bernardo, J.M. et al., 1992. On the development of reference priors. *Bayesian statistics*, 4, p.35–60.

Berry, D.A., Chaloner, K.M. & Geweke, J.K., 1995. *Bayesian Analysis in Statistics and Econometrics: Essays in Honor of Arnold Zellner* 1st ed., Wiley-Interscience.

Bertsekas, D.P. & Tsitsiklis, J.N., 2008. *Introduction to Probability* 2nd ed., Athena Scientific.

Billingsley, P., 1995. *Probability and Measure, 3rd Edition* 3rd ed., Wiley-Interscience.

Birnbaum, Z.W. & Saunders, S.C., 1969. A New Family of Life Distributions. *Journal of Applied Probability*, 6(2), p.319-327.

Björck, A., 1996. *Numerical Methods for Least Squares Problems* 1st ed., SIAM: Society for Industrial and Applied Mathematics.

Bowman, K.O. & Shenton, L.R., 1988. *Properties of estimators for the gamma distribution*, CRC Press.

Brown, L.D., Cai, T.T. & DasGupta, A., 2001. Interval estimation for a binomial proportion. *Statistical Science*, p.101–117.

Christensen, R. & Huffman, M.D., 1985. Bayesian Point Estimation Using the Predictive Distribution. *The American Statistician*, 39(4), p.319-321.

Cohen, 1991. *Truncated and Censored Samples* 1st ed., CRC Press.

Collani, E.V. & Dräger, K., 2001. *Binomial distribution handbook for scientists and engineers*, Birkhäuser.

Congdon, P., 2007. *Bayesian Statistical Modelling* 2nd ed., Wiley.

Cozman, F. & Krotkov, E., 1997. Truncated Gaussians as Tolerance Sets.

Crow, E.L. & Shimizu, K., 1988. *Lognormal distributions*, CRC Press.

Dekking, F.M. et al., 2007. *A Modern Introduction to Probability and Statistics: Understanding Why and How*, Springer.

Fink, D., 1997. A compendium of conjugate priors. *See http://www. people. cornell. edu/pages/df36/CONJINTRnew% 20TEX. pdf*, p.46.

Georges, P. et al., 2001. Multivariate Survival Modelling: A Unified Approach with Copulas. *SSRN eLibrary*.

Gupta and Nadarajah, 2004. *Handbook of beta distribution and its applications*, CRC Press.

Gupta, P.L., Gupta, R.C. & Tripathi, R.C., 1997. On the monotonic properties of discrete failure rates. *Journal of Statistical Planning and Inference*, 65(2), p.255-268.

Haight, F.A., 1967. *Handbook of the Poisson distribution*, New York,: Wiley.

Hastings, N.A.J., Peacock, B. & Evans, M., 2000. *Statistical Distributions, 3rd Edition* 3rd ed., John Wiley & Sons Inc.

Jiang, R. & Murthy, D.N.P., 1996. A mixture model involving three Weibull distributions. In *Proceedings of the Second Australia–Japan Workshop on Stochastic Models in Engineering, Technology and Management*. Gold Coast, Australia, pp. 260-270.

Jiang, R. & Murthy, D.N.P., 1998. Mixture of Weibull distributions - parametric characterization of failure rate function. *Applied Stochastic Models and Data Analysis*, (14), p.47-65.

Jiang, R. & Murthy, D.N.P., 1995. Modeling Failure-Data by Mixture of2 Weibull Distributions : A Graphical Approach. *IEEE Transactions on Reliability*, 44, p.477-488.

Jiang, R. & Murthy, D.N.P., 1999. The exponentiated Weibull family: a graphical approach. *Reliability, IEEE Transactions on*, 48(1), p.68-72.

Johnson, N.L., Kemp, A.W. & Kotz, S., 2005. *Univariate Discrete Distributions* 3rd ed., Wiley-Interscience.

Johnson, N.L., Kotz, S. & Balakrishnan, N., 1994. *Continuous Univariate Distributions, Vol. 1* 2nd ed., Wiley-Interscience.

Johnson, N.L., Kotz, S. & Balakrishnan, N., 1995. *Continuous Univariate Distributions, Vol. 2* 2nd ed., Wiley-Interscience.

Johnson, N.L., Kotz, S. & Balakrishnan, N., 1997. *Discrete Multivariate Distributions* 1st ed., Wiley-Interscience.

Kimball, B.F., 1960. On the Choice of Plotting Positions on Probability Paper. *Journal of the American Statistical Association*, 55(291), p.546-560.

Kleiber, C. & Kotz, S., 2003. *Statistical Size Distributions in Economics and Actuarial Sciences* 1st ed., Wiley-Interscience.

Klein, J.P. & Moeschberger, M.L., 2003. *Survival analysis: techniques for censored and truncated data*, Springer.

Kotz, S., Balakrishnan, N. & Johnson, N.L., 2000. *Continuous Multivariate Distributions, Volume 1, Models and Applications, 2nd Edition* 2nd ed., Wiley-Interscience.

Kotz, S. & Dorp, J.R. van, 2004. *Beyond Beta: Other Continuous Families Of Distributions With Bounded Support And Applications*, World Scientific Publishing Company.

Kundu, D., Kannan, N. & Balakrishnan, N., 2008. On the hazard function of Birnbaum-Saunders distribution and associated inference. *Comput. Stat. Data Anal.*, 52(5), p.2692-2702.

Lai, C.D., Xie, M. & Murthy, D.N.P., 2003. A modified Weibull distribution. *IEEE Transactions on Reliability*, 52(1), p.33-37.

Lai, C.-D. & Xie, M., 2006. *Stochastic Ageing and Dependence for Reliability* 1st ed., Springer.

Lawless, J.F., 2002. *Statistical Models and Methods for Lifetime Data* 2nd ed., Wiley-Interscience.

Leemis, L.M. & McQueston, J.T., 2008. Univariate distribution relationships. *The American Statistician*, 62(1), p.45–53.

Leipnik, R.B., 1991. On Lognormal Random Variables: I-the Characteristic Function. *The ANZIAM Journal*, 32(03), p.327-347.

Lemonte, A.J., Cribari-Neto, F. & Vasconcellos, K.L.P., 2007. Improved statistical inference for the two-parameter Birnbaum-Saunders distribution. *Computational Statistics & Data Analysis*, 51(9), p.4656-4681.

Limpert, E., Stahel, W. & Abbt, M., 2001. Log-normal Distributions across the Sciences: Keys and Clues. *BioScience*, 51(5), p.352, 341.

MacKay, D.J.C. & Petoy, L.C.B., 1995. A hierarchical Dirichlet language model. *Natural language engineering*.

Manzini, R. et al., 2009. *Maintenance for Industrial Systems* 1st ed., Springer.

Martz, H.F. & Waller, R., 1982. *Bayesian reliability analysis*, JOHN WILEY & SONS, INC, 605 THIRD AVE, NEW YORK, NY 10158.

Meeker, W.Q. & Escobar, L.A., 1998. *Statistical Methods for Reliability Data* 1st ed., Wiley-Interscience.

Modarres, M., Kaminskiy, M. & Krivtsov, V., 1999. *Reliability engineering and risk analysis*, CRC Press.

Murthy, D.N.P., Xie, M. & Jiang, R., 2003. *Weibull Models* 1st ed., Wiley-Interscience.

Nelson, W.B., 1990. *Accelerated Testing: Statistical Models, Test Plans, and Data Analysis*, Wiley-Interscience.

Nelson, W.B., 1982. *Applied Life Data Analysis*, Wiley-Interscience.

Novosyolov, A., 2006. The sum of dependent normal variables may be not normal. *http://risktheory.ru/papers/sumOfDep.pdf*.

Patel, J.K. & Read, C.B., 1996. *Handbook of the Normal Distribution* 2nd ed., CRC.

Pham, H., 2006. *Springer Handbook of Engineering Statistics* 1st ed., Springer.

Provan, J.W., 1987. Probabilistic approaches to the material-related reliability of fracture-sensitive structures. *Probabilistic fracture mechanics and reliability(A 87-35286 15-38). Dordrecht, Martinus Nijhoff Publishers, 1987,*, p.1–45.

Rao, C.R. & Toutenburg, H., 1999. *Linear Models: Least Squares and Alternatives* 2nd ed., Springer.

Rausand, M. & Høyland, A., 2004. *System reliability theory*, Wiley-IEEE.

Rencher, A.C., 1997. *Multivariate Statistical Inference and Applications, Volume 2, Methods of Multivariate Analysis* Har/Dis., Wiley-Interscience.

Rinne, H., 2008. *The Weibull Distribution: A Handbook* 1st ed., Chapman & Hall/CRC.

Schneider, H., 1986. *Truncated and censored samples from normal populations*, M. Dekker.

Simon, M.K., 2006. *Probability Distributions Involving Gaussian Random Variables: A Handbook for Engineers and Scientists*, Springer.

Singpurwalla, N.D., 2006. *Reliability and Risk: A Bayesian Perspective* 1st ed., Wiley.

Sison, C.P. & Glaz, J., 1995. Simultaneous Confidence Intervals and Sample Size Determination for Multinomial Proportions. *Journal of the American Statistical Association*, 90(429).

Tong, Y.L., 1990. *The Multivariate Normal Distribution*, Springer.

Xie, M., Gaudoin, O. & Bracquemond, C., 2002. Redefining Failure Rate Function for Discrete Distributions. *International Journal of Reliability, Quality & Safety Engineering*, 9(3), p.275.

Xie, M., Goh, T.N. & Tang, Y., 2004. On changing points of mean residual life and failure rate function for some generalized Weibull distributions. *Reliability Engineering and System Safety*, 84(3), p.293–299.

Xie, M., Tang, Y. & Goh, T.N., 2002. A modified Weibull extension with bathtub-shaped failure rate function. *Reliability Engineering and System Safety*, 76(3), p.279–285.

Yang and Berger, 1998. A Catalog of Noninformative Priors. http://www.stats.org.uk/priors/noninformative/YangBerger1998.pdf, Last checked on Nov. 16, 2018.

Yang, K. et al., 2004. *Multivariate Statistical Methods in Quality Management* 1st ed., McGraw-Hill Professional.

Yang, R. & Berger, J.O., 1994. Estimation of a Covariance Matrix Using the Reference Prior. *The Annals of Statistics*, 22(3), p.1195-1211.

Zhou, X.H. & Gao, S., 1997. Confidence intervals for the log-normal mean. *Statistics in medicine*, 16(7), p.783–790.